Send all inquiries to:
GLENCOE DIVISION
Macmillan/McGraw-Hill
3008 W. Willow Knolls Drive
Peoria, IL 61615

Printed in the United States of America.

Orders and requests for information about cost and availability of yearbooks should be addressed to the company.

Request to quote portions of yearbooks should be addressed to the Secretary, Council on Technology Education, in care of the publisher, for forwarding to the current Secretary.

This publication is available in microform.

University Microfilms International
300 North Zeeb Road
Dept. P.R.
Ann Arbor, MI 48106

ISBN 0-02-677141-1

CONSTRUCTION IN TECHNOLOGY EDUCATION

EDITORS

Jack W. Wescott, Ph.D.
Assistant Professor
Department of Industry and Technology
Ball State University

Richard M. Henak, Ed.D.
Professor
Department of Industry and Technology
Ball State University

43rd Yearbook, 1994

Council on Technology
Teacher Education

GLENCOE

Macmillan/McGraw-Hill

New York, New York   Columbus, Ohio   Mission Hills, California   Peoria, Illinois

FOREWORD

Technology teacher education is facing an exciting, yet challenging time in the history of the profession. One of the tasks of technology teacher education is to redirect the curriculum to reflect contemporary technology education.

This yearbook is the fourth and last of a series that deals with the general implementation of and instructional strategies for the content organizers of communication, transportation, manufacturing, and construction. Previous yearbooks, *Implementing Technology Education* (1986), and *Instruction Strategies for Technology Education* (1988) established the setting for this series of yearbooks.

Construction has and will always remain important in the evolution of technology. The editors have noted that:

> The construction industry is one of the largest industries in the world, but we have reason for concern about its future. In light of the industry's significance and uniqueness, and its increasing complexity, few people understand the intricacy of how the industry functions and the opportunities it provides. Construction concepts must be introduced in public education and taught in ways that help students develop better understandings of the world in which we live and the society to which the student will contribute as an adult. (Preface)

A great need exists for a yearbook that will serve as a resource for the development of quality construction technology education programs at the elementary, secondary, and university levels.

The first three chapters highlight the rationale and structure of content for studying construction; the past, present, and future of construction technology; and the impacts of construction technology. The next four chapters provide direction for implementing construction at the elementary, secondary levels, and teacher education programs. Chapters eight, nine, and ten identify the instructional methods, learning environments, and assessment techniques for construction in technology education. The final two chapters provide a perspective of construction technology in the third world and a summary.

On behalf of the CTTE members, I commend Drs. Jack Wescott and Richard Henak for doing an excellent job in editing this valuable yearbook. Appreciation is also expressed to Glencoe Division of Macmillan/McGraw-Hill for their continued support of the CTTE yearbooks.

Everett N. Israel
President, Council on Technology Teacher Education

YEARBOOK PLANNING COMMITTEE

Terms Expiring 1994

Anthony E. Schwaller
St. Cloud State University
Robert E. Wenig
North Carolina State University

Terms Expiring 1995

Richard M. Henak
Ball State University
James E. LaPorte
Virginia Polytechnic Institute and State University

Terms Expiring 1996

M. Roger Betts
University of Northern Iowa
Jane A. Liedtke
Illinois State University

Terms Expiring 1997

Donald P. Lauda
University of California-Long Beach
G. Eugene Martin
Southwest Texas State University

Terms Expiring 1998

Stanley A. Komacek
California University of Pennsylvania
John R. Wright
Central Connecticut State University

Chairperson

R. Thomas Wright
Ball State University

OFFICERS OF THE COUNCIL

President

Everett N. Israel
Department of Industrial Technology
Eastern Michigan University
Ypsilanti, MI 48197

Vice-President

Richard M. Henak
Department of Industry & Technology
Ball State University
Muncie, IN 47306

Recording Secretary

Betty L. Rider
Assistant Principal
Rutheford B. Hayes High School
Deleware, OH 43015

Membership Secretary

Gerald L. Jennings
Department of Business and Industrial Education
Eastern Michigan University
Ypsilanti, MI 48197

Treasurer

Michael D. Wright
Technology Education
Montana State University
Bozeman, MT 59717

Past President

R. Thomas Wright
Department of Industry and Technology
Ball State University
Muncie, IN 47306

YEARBOOK PROPOSALS

Each year, at the ITEA International Conference, the CTTE Yearbook Committee reviews the progress of yearbooks in preparation and evaluates proposals for additional yearbooks. Any member is welcome to submit a yearbook proposal. It should be written in sufficient detail for the committee to be able to understand the proposed substance and format. Fifteen copies of the proposal should be sent to the committee chairperson by February 1 of the year in which the conference is held. Below are the criteria employed by the committee in making yearbook selections.

CTTE Yearbook Committee

CTTE Yearbook Guidelines

A. Purpose:

The CTTE Yearbook Series is intended as a vehicle for communicating education subject matter in a structured, formal series that does not duplicate commercial textbook publishing activities.

B. Yearbook topic selection criteria:

An appropriate yearbook topic should:

1. Make a direct contribution to the understanding and improvement of technology teacher education.
2. Add to the accumulated body of knowledge of the field.
3. Not duplicate publishing activities of commercial publishers or other professional groups.
4. Provide a balanced view of the theme and not promote a single individual's or institution's philosophy or practices.
5. Actively seek to upgrade and modernize professional practice in technology teacher education.
6. Lend itself to team authorship as opposed to single authorship.

Proper yearbook themes *may* also be structured to:

1. Discuss and critique points of view which have gained a degree of acceptance by the profession.
2. Raise controversial questions in an effort to obtain a national hearing.
3. Consider and evaluate a variety of seemingly conflicting trends and statements emanating from several sources.

C. The yearbook proposal:

1. The Yearbook Proposal should provide adequate detail for the Yearbook Planning Committee to evaluate its merits.
2. The Yearbook Proposal should include:
   a. An introduction to the topic
   b. A listing of chapter titles
   c. A brief description of the content or purpose of each chapter
   d. A tentative list of authors for the various chapters
   e. An estimate of the length of each chapter

PREVIOUSLY PUBLISHED YEARBOOKS

*1. *Inventory Analysis of Industrial Arts Teacher Education Facilities, Personnel and Programs,* 1952.
*2. *Who's Who in Industrial Arts Teacher Education,* 1953.
*3. *Some Components of Current Leadership: Techniques of Selection and Guidance of Graduate Students; An Analysis of Textbook Emphases;* 1954, three studies.
*4. *Superior Practices in Industrial Arts Teacher Education,* 1955.
*5. *Problems and Issues in Industrial Arts Teacher Education,* 1956.
*6. *A Sourcebook of Reading in Education for Use in Industrial Arts and Industrial Arts Teacher Education,* 1957.
*7. *The Accreditation of Industrial Arts Teacher Education,* 1958.
*8. *Planning Industrial Arts Facilities,* 1959. Ralph K. Nair, ed.
*9. *Research in Industrial Arts Education,* 1960. Raymond Van Tassel, ed.
*10. *Graduate Study in Industrial Arts,* 1961. R. P. Norman and R. C. Bohn, eds.
*11. *Essentials of Preservice Preparation,* 1962. Donald G. Lux, ed.
*12. *Action and Thought in Industrial Arts Education,* 1963. E. A. T. Svendsen, ed.
*13. *Classroom Research in Industrial Arts,* 1964. Charles B. Porter, ed.
*14. *Approaches and Procedures in Industrial Arts,* 1965. G. S. Wall, ed.
*15. *Status of Research in Industrial Arts,* 1966. John D. Rowlett, ed.
*16. *Evaluation Guidelines for Contemporary Industrial Arts Programs,* 1967. Lloyd P. Nelson and William T. Sargent, eds.
*17. *A Historical Perspective of Industry,* 1968. Joseph F. Luetkemeyer Jr., ed.
*18. *Industrial Technology Education,* 1969. C. Thomas Dean and N. A. Hauer, eds. *Who's Who in Industrial Arts Teacher Education,* 1969. John M. Pollock and Charles A. Bunten, eds.
*19. *Industrial Arts for Disadvantaged Youth,* 1970. Ralph O. Gallington, ed.
*20. *Components of Teacher Education,* 1971. W. E. Ray and J. Streichler, eds.
*21. *Industrial Arts for the Early Adolescent,* 1972. Daniel J. Householder, ed.
*22. *Industrial Arts in Senior High Schools,* 1973. Rutherford E. Lockette, ed.
*23. *Industrial Arts for the Elementary School,* 1974. Robert G. Thrower and Robert D. Weber, eds.
*24. *A Guide to the Planning of Industrial Arts Facilities,* 1975. D. E. Moon, ed.
*25. *Future Alternatives for Industrial Arts,* 1976. Lee H. Smalley, ed.
*26. *Competency-Based Industrial Arts Teacher Education,* 1977. Jack C. Brueckman and Stanley E. Brooks, eds.
*27. *Industrial Arts in the Open Access Curriculum,* 1978. L. D. Anderson, ed.
*28. *Industrial Arts Education: Retrospect, Prospect,* 1979. G. Eugene Martin, ed.
*29. *Technology and Society: Interfaces with Industrial Arts,* 1980. Herbert A. Anderson and M. James Benson, eds.
*30. *An Interpretive History of Industrial Arts,* 1981. Richard Barella and Thomas Wright, eds.
*31. *The Contributions of Industrial Arts to Selected Areas of Education,* 1982. Donald Maley and Kendall N. Starkweather, eds.
*32. *The Dynamics of Creative Leadership for Industrial Arts Education,* 1983. Robert E. Wenig and John I. Mathews, eds.
*33. *Affective Learning in Industrial Arts,* 1984. Gerald L. Jennings, ed.
*34. *Perceptual and Psychomotor Learning in Industrial Arts Education,* 1985. John M. Shemick, ed.
*35. *Implementing Technology Education,* 1986. Ronald E. Jones and John R. Wright, eds.
*36. *Conducting Technical Research,* 1987. Everett N. Israel and R. Thomas Wright, eds.
*37. *Instructional Strategies for Technology Education,* 1988. William H. Kemp and Anthony E. Schwaller, eds.
38. *Technology Student Organizations,* 1989. M. Roger Betts and Arvid W. Van Dyke, eds.
39. *Communication in Technology Education,* 1990. Jane A. Liedtke, ed.
40. *Technological Literacy,* 1991. Michael J. Dyrenfurth and Michael R. Kozak, eds.
41. *Transportation in Technology Education,* 1992. John R. Wright and Stanley Komacek, eds.
42. *Manufacturing in Technology Education,* 1993. Seymour and Shackelford, eds.

---

*Out-of-print yearbooks can be obtained in microfilm and in Xerox copies. For information on price and delivery, write to Xerox University Microfilms, 300 North Zeeb Road, Ann Arbor, Michigan, 48106.

CONTENTS

=== **Chapter 1**

Richard M. Henak, Ed.D.
Ball State University

=== **Chapter 2**

David A. Ross, Ed.D.
Georgia Southern University

=== **Chapter 3**

Peter H. Wright, Ed.D.
Indiana State University

=== **Chapter 4**

James J. Kirkwood, Ph.D.
Ball State University

PREFACE

The construction industry is one of the largest industries in the world, but we have reason to be concerned about its future. In light of the industry's significance, its uniqueness, and its increasing complexity, few people understand the intricacy of how the industry functions and the opportunities it provides. Construction concepts must be introduced in public education and taught in ways that help students develop better understandings of the world in which we live and the society to which the student will contribute as an adult.

Therefore, a great need exists for an activity-oriented curriculum that develops problem-solving and decision-making skills, and expands basic construction literacy. A well-designed construction technology education program can provide young people with an understanding of the construction industry and help them to participate effectively in our society.

The major purpose of this yearbook is to improve technology education. It should be noted that it was never intended to be an immediate and ultimate answer for questions pertaining to the profession. However, the chapter titles were selected in order to provide the reader with assistance in the following three areas: 1) the rationale and content for construction in technology education, 2) the selection of instructional methods, and 3) guidelines for implementation.

Finally, it is our hope that this yearbook can serve as a point of reference for present and future discussions of construction in technology education. Furthermore it is hoped that this publication will make a valuable contribution to the technology education profession at all levels, and especially to the many young individuals contemplating entering the profession now and in the future.

Jack W. Wescott

Richard M. Henak

ACKNOWLEDGMENTS

We wish to recognize the many individuals whose contributions have made this yearbook a reality. In particular, we thank the chapter authors for their time, professional commitment, and dedication to construction in technology education. Without the dedication and commitment of the authors, this yearbook would not have been a reality.

We would also like to acknowledge the members of the Yearbook Committee for their understanding and direction during the development of this yearbook. The committee's input was extremely helpful throughout the process.

It is also important to recognize the support we received from our colleagues in the Department of Industry and Technology at Ball State University.

A special thanks also is extended to the Glencoe Division of Macmillan/McGraw-Hill for their continued support of the yearbook series.

Finally, a very special thank you to our families for their patience and support during the three years of this undertaking. It is our sincere hope that this yearbook will assist the profession in the development and implementation of quality construction programs in technology education.

Jack W. Wescott

Richard M. Henak

CHAPTER 1

Rationale and Structure of Content for Construction in Technology Education

Richard M. Henak, Ed.D.,
Department of Industry and Technology
Ball State University, Muncie, IN

From the Great Wall of China to the Forbidden City, from the Vatican Palace to the Crystal Palace, from the Statue of Liberty and Mount Rushmore to the Panama Canal and the Trans-Siberian Railway, they bear witness to the visionaries who stretched the limits of imagination to conceive them and the builders who bent the laws of nature to construct them. (Hawkes, 1990, jacket)

For this chapter, *construction technology is the study of the efficient practice of using production and management processes to transform materials and assemble components into buildings, and heavy industrial and civil structures that are built on site.* The construction industry provides the built environment for its users. Cities, streets, houses, schools, parks, skyscrapers, bridges, and barns affect our actions. In many ways, they decide how we live. The built environment provides the surroundings in which we walk, sleep, eat, travel, work, or play. There is little doubt we are affected by it. Especially, if the effect is pleasant and inspires us; or, when it is dreary and doesn't meet our expectations. All players in the construction drama are needed, but it is the users who all others serve.

Take a moment to observe your surroundings. Examples of outputs from the construction system are everywhere. Structures shelter us and our possessions from the forces of nature and allow us to live in any climate. Many things are healthier, more comfortable, and more productive when protected from the natural elements — snow, cold, heat, rain, and wind. Barns provide shelter for livestock; greenhouses protect tender plants; people build houses so they are secure, warm and dry; a water tank keeps water clean and available; and warehouses protect products of all kinds.

Industrial buildings such as small factories and warehouses are used to produce and store materials and products used all over the world. Industrial complexes are built to produce large quantities of products such as steel, paper, cement, fertilizers, chemicals, fuels, electrical energy, and ships.

Transportation systems move people, products, and energy to places where they are needed. Highways make it easy to move people and products over long distances and to isolated areas. Pipeline systems, that consist of gathering lines, processing, and distribution facilities, supply fresh water and fuels, and carry away wastes. These systems collect, process, and transport solids (in the form of a slurry), liquids, and gases. The materials are moved between buildings; across continents; and over, through, and under ground and water. Electrical transmission systems use heavy metal cables buried deep in the ground, placed under water, or suspended from towers high above the ground to distribute electrical energy. A network of cables, pipes, and tunnels that carry water, steam, sewage, information, electricity, and natural gas exists beneath the streets of a modern city. These systems are so quiet and work so well that we tend to forget about them. Modern utilities and services are possible because of construction technology.

Communications systems include buildings that house transmitters and receivers, systems of towers that hold antennas high above the ground, and satellites that are placed in orbit. All are used to communicate information to and from space, and to every point on the surface of the earth. Even road signs are designed and constructed to provide motorists with essential information.

A RATIONALE FOR THE STUDY OF CONSTRUCTION TECHNOLOGY

Deciding what to teach and how to teach it are critical questions asked by technology educators. Teachers of construction must decide on objectives, select texts and materials, design instructional strategies, prepare learning activities, and plan assessment methods. Although making these decisions seem simple, it has proven to be difficult and demanding work.

Developing a course rationale helps in three important ways. First, participation in the process builds our understanding and clarifies our thinking about teaching. Second, the rationale serves as a guide in the search for alternative solutions and provides a basis for making decisions related to the teaching/learning process. Finally, we can effectively respond to students when they ask, "Why do we have to learn this?"

Developing A Rationale

In his book *Basic Principles of Curriculum and Instruction*, Tyler (1975) proposed a model for preparing a philosophy. Tyler's model suggests that our philosophy of education consists of our beliefs regarding the students, society, subject matter, purposes of education, and learning theory.

Course writers draw upon their philosophies to build a course rationale. Figure 1–1 illustrates how a rationale emerges from a philosophy. It has all of the elements of a philosophy, but is less than the philosophy of education because it is focused on a single course with a specific and limited scope and purpose. Even if the same person wrote both rationales, a rationale for a course designed to develop the workforce is different from a rationale for a course designed to develop technological literacy.

Characteristics and Needs of Students. Because advertisers often understand the wants and needs of potential buyers better than the customers themselves, they are able to present products in appealing ways. An effective television commercial must be interesting enough to get potential buyers to watch it, to learn about the product, and to purchase the product. Clearly, television commercials would be less successful if advertisers did not understand the characteristics of the audience.

In this respect, teaching is similar to television advertising. We have a product, in the form of instruction, that consists of knowledge, skills, and feelings that we want our students to experience. The better we understand our students, the better we can present the learning experience in an appealing way. If we are successful, our students will pay attention, achieve the objectives, and apply what they learn.

In the last 25 years, considerable research has been done on the different ways people prefer to learn (Henak, 1992, p. 23). Multiple learning experiences should be designed so students with different styles of learning modalities and talents have opportunities to succeed.

Students also have needs. From one-third to one-half of high school students are enrolled in what is often termed a general education track which stresses neither academic nor workplace preparation. It is usually an instructional "hodgepodge" of courses which prepare students neither for continuing their education nor for employment. The writers of the Secretary's Commission on Achieving Necessary Skills (SCANS, 1992) found that the high school diploma has lost its value. They conclude that, "in many places today . . . a high school diploma is little more than a certificate of attendance" (pp. 6–7).

Students need accurate information regarding career opportunities. Construction provides many interesting challenges and rewards for capable young people who wish to work at any career level. R. "Mac" Sullivan, Jr.

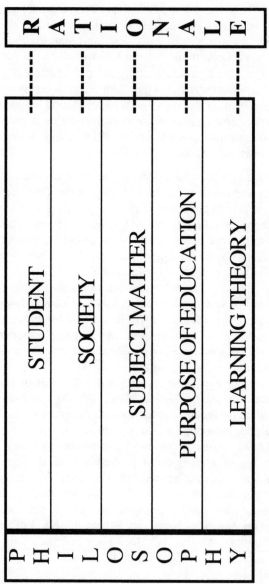

Figure 1–1: Relationship of a philosophy of education to a rationale.

(1989, p. 18), Secretary of the Construction Industry Workforce Foundation and Chairman of its Image Committee, discovered constructors feel the construction industry: (1) offers many opportunities for earning while learning, (2) extends an opportunity for a fast-track career path, (3) presents many interesting challenges, (4) provides a lot of hands-on satisfaction, (5) epitomizes the "small business entrepreneurial spirit of America," and (6) yields great personal pride in those who see structures they built throughout their community.

Characteristics and Needs of Society.   Our beliefs about society can be categorized into the characteristics of society and the needs of society. In his book, *What's Worth Teaching?*, Marion Brady (1989) states,

> We look at reality not directly but through a "lens" created by our culture. That lens is ground to a prescription unlike any other. There are distortions in the glass. It is tinted a distinctive hue. Certain areas are cloudy, others opaque. Specks of dirt and dust adhere to the lens. (p. 14)

In varying degrees, people view society in much the same way. A clearer more complete view of society can be gained by studying construction technology. The characteristics of our society are shaped in many ways by construction technology. The world is not static. Information is expanding at an ever-accelerating rate and the rapidity of social change has never been greater. Each curriculum area has the fortune and misfortune of growth. Knowledge has replaced brawn, land, and other natural resources as the leading factor in determining our gross national product. Learning-to-learn has become an imperative in technology education so that students can renew and sustain their education throughout life. But developing learning skills alone is not enough. The subject matter of construction technology must be selected for its generalizability in human experiences. Traditional discipline-bound, fact-laden, point-of-practice construction courses are too narrow in scope to accomplish this.

Society is both served and controlled by construction technology. Products of construction technology impact our lives. Construction technology was used to produce the most awe-inspiring feats of human creation. On every continent there are superlative achievements of architecture and engineering. These accomplishments are the largest, longest, highest, and most ambitious structures ever made by human hands.

In addition to changing our landscape, the construction industry drives our economy. When the construction industry is growing, industrial production is busy producing glass, steel, copper, aluminum, natural and synthetic fibers, aggregates, cement, furnishings, computers, pick-up trucks, and heavy equipment (Robins, 1993).

The initial taxonomy described in *A Rationale and Structure for Industrial Arts Subject Matter* (Towers, Lux, & Ray, 1966) and later refined for *The World of Construction* (Lux & Ray, 1970) was clearly focused on the subject matter derived from an analysis of the construction industry. Curriculum was designed to have students experience construction management, personnel, process, and community planning technology used to build the smallest and simplest structures to the largest and most complex. People and their relationship to the challenges, thought processes, tasks, and work with construction projects should be emphasized rather than the intellectual mastery of construction information. More students would benefit from a curriculum with a broader vision and one that reflects the expanding richness of construction technology.

Society also has needs. The construction industry is one of the largest industries in the world, but in light of the industry's significance, uniqueness, and increasing complexity, few people understand the intricacy of how the industry functions, its impact on our nation, and the opportunities it provides. Construction concepts need to be introduced in public education and taught using strategies that help society develop better appreciations and understandings of the world in which we live and the society to which the student will contribute as an adult.

Jackson (1992a), President of Associated General Contractors of America (AGC of A), states, "every center of production in America will be productive when the construction industry is even close to working at capacity" (p. 5). The amounts of dollars invested in construction since 1982 (in constant 1987 dollars) is shown in Figure 1–2.

Subject Matter.   Teachers differ in how they approach the identification of subject matter. Some think the primary source for content is an analysis of literature. Others believe learners and their needs provide the most useful source. The message to teachers from the SCANS report entitled, *Learning a Living: A Blueprint for High Performance* (SCANS, 1992), is to, "Look beyond your discipline and your classroom to the other courses your students take, to your community, and to the lives of your students outside school. Help students connect what they learn in class to the world outside" (p. xiii).

Construction experiences provide students with opportunities to use this generalizable subject matter to design, plan, and construct a built environment that offers a higher quality of life, a favorable environment for human survival, sustainable energy resources, physical and mental well-being, and longevity.

Year	Dollars*
1982	$297.8 Billions
1983	$332.6 Billions
1984	$382.4 Billions
1985	$402.0 Billions
1986	$421.4 Billions
1987	$419.4 Billions
1988	$415.1 Billions
1989	$409.5 Billions
1990	$397.5 Billions
1991	$358.5 Billions
1992	$378.5 Billions
* - In constant 1987 dollars	

Figure 1–2:   Investments in construction since 1982 (Jackson, 1992a).

A second fundamental consideration relates to the nature of the content. Should the content be subject-oriented, product-oriented, or process-oriented? In the subject-oriented approach, attention is given to developing an understanding of the fundamental structure of the subject matter and to how the subject matter can be organized so it is easier to learn and use in real-life situations. Users of the product-oriented approach direct their attention toward the "destination" rather than the "journey" and on "what" is produced. In the process-oriented approach, instruction is focused on the "journey" and on "how" work is accomplished, problems are solved, and decisions are made.

A process-oriented construction technology subject matter (Indiana Industrial Technology Education Curriculum, 1992a, b, & c; Henak, 1993) reflects current project delivery processes and approaches, settings, and relationships. Actions of owners who initiate projects, designers who convert the needs of owners into contract documents, constructors who convert construction documents and resources into structures, and owners who use structures are potential roles students can play. Curriculum based on materials or skills is often fragmented. Fragmenting the construction curriculum destroys the unity of the process used to design, build, and use the built environment.

Comprehensive subject matter-oriented construction programs include effective management and efficient practices used by owners, designers, and

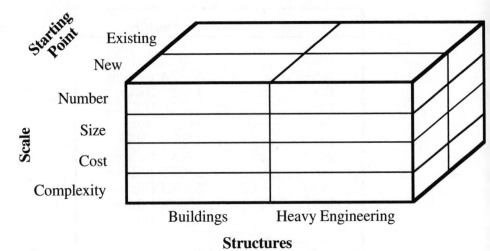

Structures

Figure 1–3:   Types, scales, and starting points of construction projects.

constructors to design, build, and use structures. The subject matter is carefully analyzed and organized into carefully structured units. *The World of Construction* (Lux & Ray, 1970) was a thoroughly done example of a subject-oriented construction book. The conceptual model in Figure 1–3 illustrates three ways to broaden the customary way structures are defined. They are: (1) the structure types, (2) the scale of construction projects, and (3) the starting point for a project.

Finally, who decides on the course content? Generally, the teacher determines what is to be taught, but that is not always true. For example, the AGC of A specifies the content and methods for teaching their courses Partnering, Total Quality Management, Risk Allocation, and numerous other training opportunities they sponsor. State departments of education and school administrators may also specify the course content. At the other extreme, learners may identify some, or all, of the content for independent study courses.

Purpose of Education.   Beliefs about the purpose for courses directly influence planning. It is essential that beliefs about the purpose of education be clearly understood. These subjective factors significantly influence one's ultimate satisfaction or dissatisfaction with their teaching as well as what actually happens in the classroom. L. Dee Fink (1984) asked a group of first year teachers to complete the following sentence: "The most important thing I want to do for students is . . ." (p. 62). Their responses varied greatly, but after an analysis was completed, five basic lines of thinking were revealed. They were to: (1) promote general intellectual growth, (2) teach mastery of the subject matter or discipline, (3) develop application skills for a vocation or for life, (4) engage or develop students' feelings, and (5) prepare students for further learning. Dr. Fink concluded that, "The majority of the new teachers were trying not just to teach a specific subject matter but to achieve some greater good, some general intellectual growth or to lay the groundwork for future learning" (p. 63).

Kozak and Robb (1991) wrote that, traditionally, American people have "regarded education as a means for improving themselves and their society" (p. 34). They concluded that the following two purposes were still relevant:

1. the set of "seven cardinal principles" (Commission on Reorganization of Secondary Education, 1918) — health, command of fundamental processes, worthy home membership, vocational competence, effective citizenship, worthy use of leisure, and ethical character; and

2. the four headings for categories of objectives — self-realization, human relationships, economic efficiency, and civic responsibility (Educational Policies Commission, 1938, 1961).

Technological literacy assumes increasing importance as technology instructors seek to prepare young persons to meet the challenges of the future. Dyrenfurth and Kozak (1991, p. 2), the editors of *Technological Literacy* reviewed literature in the fields of social policy/commentary, technology, science, and liberal arts/humanities and concluded the roots of technological literacy could be found in the categories of:

1. *democratic needs* - If we live and are to function effectively in a democracy infused with technology, we must understand technology.

2. *nature of life in society* - We must be capable with and understand technology to participate or even survive in a rapidly changing technological world.

3. *dehumanization/humanization* - Technology can and does dehumanize people by putting demands on us we cannot control.

4. *new liberal arts directions* - Understanding technology in order to determine the ethical issues involved with technology and education for it.

5. *nature of jobs/competitiveness/workforce literacy* - An understanding of technology is essential if we are to achieve the goal of a high performance economy characterized by "high skills, high wages, and full employment" (SCANS, 1991, p. 4).

6. *technology as a discipline* - Technology is being promoted as a subject in its own right.

Dyrenfurth (1991), described technological literacy as "a multi-dimensional term that necessarily includes *the ability to use* technology (practical dimension), the ability to understand the issues raised by or use of technology (civic dimension), and the appreciation for the significance of technology (cultural dimensions)" (p. 7).

Traditional construction courses focus on the skills used to draw and build houses which have limited appeal and utility. Current practices, accurate terms, typical paperwork, and industrial procedures are often used to enhance the simulation of real-life scenarios. Appropriate use of computers and calculators enhance both the students' understanding of construction technology and their mastery of basic skills. Students can engage in activities, make decisions, solve problems, and experience real-life situations that face owners, designers, and constructors. This can be enriched by asking adults outside the class to serve as owners and representatives of regulatory agencies when structures are being designed, community plans are presented, or bids are opened.

Learning Theory. Like each of the other components of a philosophy of teaching, learning theory is extremely complex. A simple way to view learning theory is to look at three of its basic elements: things we know about motivation, designing curriculum, and working with students (Zemke & Zemke, 1981).

There are two kinds of motivation. First, extrinsic variables of motivation involve the achievement of a reward or goal that may be provided by others or by the person. Secondly, intrinsic motivation is when the activity itself is rewarding. Rewarding experiences are the ones that students find meaningful and enjoyable. Robert Marzano (1992), in *A Different Kind of Classroom: Teaching with Dimensions of Learning*, states that, "Generally, we acquire and integrate knowledge because we want or need to use it" (p. 106). He goes on to list decision making, investigation, experimentation, inquiry, problem solving, and invention as "common ways we use knowledge meaningfully" (p. 106). The nature of the curriculum should be such that it will engage all students in active, meaningful participation in real experiences.

Long-term, student-directed, applications-based activities are designed to provide purpose and direction to learning and tend to involve students (Marzano, 1992, p. 124). These strategies make students active participants in their own learning rather than passive observers and receivers of knowledge. Student-centered, real-world scenarios enhance interest. Life provides us with a model setting for learning of all kinds. Realistic settings add meaning and facilitate classroom learning when students respond to daily problems, issues, and opportunities. You can include conventional terminology, use job titles, play work roles, apply management procedures, employ typical documents, and establish the relationship among owners, designers, constructors, and suppliers.

Unique characteristics of the construction industry can be simulated. Actions of owners who initiate projects, designers who convert the needs of owners into contract documents, constructors who convert construction documents and resources into structures, and owners who use structures are some typical roles. People and their relationship to the challenges, thought processes, tasks, and work with the construction project should be emphasized rather than the intellectual mastery of information about construction.

Instruction consists of leading students through learning experiences to help them grasp, transform, and transfer what is learned into personal meanings. There are usually several ways to design curriculum to facilitate learning for students.

If we were asked to make cookies and given the ingredients for some cookies, no doubt we would ask what kind and ask to see a recipe. It is the knowledge of the kind of cookies and the recipe that assures us that the ingredients are in the correct quantities, that they were mixed properly, that none are left out, and they were baked correctly. Without the recipe we probably would not bother with making cookies.

Ironically, this is quite often the very thing we ask our students to do in school. Our best educators tell us that all knowledge is related, but traditional academic disciplines do not fit together in any coherent manner. Acquiring and integrating knowledge involves using what you already know to make sense out of new information, working out the kinks in the new information, and assimilating the information so that it can be used with relative ease.

Construction technology connects in many ways with all other systems of technology education and with different subject fields. Activities should be designed with active linkages between fields of knowledge. Construction projects can be designed or built with an eye to history, government regulations, values typical of owners, designers and builders; and to the role of science, mathematics, and communications. Curriculum integration is more than an interesting diversion. It is more relevant when there are connections between subjects rather than narrow isolation. Students must be given opportunities to sense this connectedness in their school experience.

CONCEPTUAL MODELS FOR THE STUDY OF CONSTRUCTION TECHNOLOGY

Experts have difficulty understanding, let alone agreeing upon, the nature of curriculum and its relationship to instruction. For the purpose of this chapter, a definition is taken from Allen A. Glatthorn's (1987) book, *Curriculum Renewal*.

> The curriculum is the plans made for guiding learning in schools, usually represented in retrievable documents of several levels of generality, and the implementation of those plans in the classroom; those experiences take place in a learning environment that also influences what is learned. (p. 1)

The curriculum includes both a description of the intent and the implementation of these plans. The nature of the intent and classroom activities can be found recorded in such materials as in course guides, syllabi, handouts, lesson plans, design briefs, instructional media, and assessment instruments.

Models For Construction Technology

In the early 1960's, industrial arts education was affected by a general thrust toward curriculum improvement. Concerned as they were with industrial processes, products, materials, and occupations, industrial arts educators were increasingly aware of the growing gap between societal realities and their representation in the total educational program. More particularly, it was quite evident that many of the traditional approaches to industrial arts education were incapable of providing students with an adequate understanding of the impact of industry upon our technological world and its population. It is generally recognized that the central question involved in bringing about a major change in industrial arts education was the question of instructional content.

Industrial Arts Curriculum Project: World of Construction (Towers, et al. 1966). The Industrial Arts Curriculum Project (IACP) was a systematic and thorough undertaking to develop a rationale and to conceptualize the structure of an organized study of industry. IACP was a joint effort by The Ohio State University and the University of Illinois with financial support coming from the U.S. Office of Education. The project staff developed a basic structure of the body of knowledge which it defined as industrial technology with two elements — construction technology and manufacturing technology.

The World of Construction (Lux & Ray, 1970) course was selected as the initial experience in the junior high school program because construction activity is more visible than manufacturing activity. Construction projects are sprinkled throughout cities, towns, and the countryside; and many students may have observed structures being built by their parents or friends. In contrast, manufacturing takes place in factories out of sight of most people, even of those who live across the street or next door to a manufacturing firm.

A task force of construction specialists provided the finite elements and subheadings which were needed to further detail the basic structure which was validated by a professional peer group and an advisory committee consisting of consultants from other disciplines, the construction industry, and professional organizations.

The next step was to develop a position paper that described the criteria used to select learning experiences from the body of knowledge. A detailed instructional handbook was developed, field tested, revised, and converted into an instructional system with an integrated package. The program was then introduced in selected experimental centers throughout the nation. In-service workshops and curriculum consultants from the project staff aided the local teachers and administrators. A thorough evaluation of the experimental programs was conducted. A refined instructional system consisting of a textbook, a student laboratory manual, teacher's guide, and suggested apparatus were provided by McKnight & McKnight Publishing Company (Lux & Ray, 1970a, 1970b, & 1970c).

The IACP staff were the first to make management and personnel technology and community planning concepts equal partners to production technology which dominated industrial arts programs. The management content included initiating the project, designing and engineering the project, and selecting a builder. Personnel technology included contracting, collective bargaining, hiring, training, working, advancing, and handling grievances with mediation, arbitration, and strikes.

The World of Construction was designed to provide a comprehensive one-year course of 180 assignments. Later versions (Lux, et al., 1982) divided the course into "Quads" so that a variety of nine-week or semester courses with 45 and 90 assignments respectively are grouped in a rational manner. It dealt with land use and the facilities constructed on a given site to meet a need. *The World of Construction* was a comprehensive and innovative course with a total instructional system for use by students in the junior high school, but was and is today taught in senior high schools. The course was designed to provide a level of conceptual development needed by citizens to understand the construction industry. Millions of students have been exposed to this activity-oriented program in the last decade. It is in its fifth edition.

The results of the work by the IACP staff has contributed greatly to the writing of construction technology books (Betts, Fannin, Hauenstein, 1976; Lux, Ray, Blankenbaker, Umstattd, 1982; Henak, 1993; Wright & Henak, 1993) and state curriculum courses such as Introduction to Construction (Indiana Industrial Technology Education Curriculum, 1992) and Exploring Construction Technology (Wood, 1987).

Jackson's Mill Industrial Arts Curriculum Theory (Snyder & Hales, 1981). In the mid-sixties, it was becoming increasingly difficult to determine which goals and activities the industrial education profession should pursue. Confusion abounds within and between industrial arts and vocational education. Leslie H. Cochran (1968) recognized the confusion in the profession and, for a dissertation, analyzed 20 contemporary, innovative programs in terms of the influences that were shaping programs at that time. His findings were reported in a popular book entitled, *Innovative Programs in Industrial Education* (1970).

Confusion prevailed and intensified. The literature in our field was replete with concerns and warnings about the direction and future of industrial arts (Snyder & Hales, 1981, p. ii). It was apparent that the time had arrived for the leaders to rededicate themselves to a common professional cause. The Jackson's Mill Curriculum Theory Symposium provided a rare opportunity for 21 dedicated industrial arts teacher educators to think cooperatively and creatively to inquire, assimilate, compromise, and come to consensus on a rationale and direction for the future of industrial arts.

A major contribution the Jackson's Mill participants made was to continue the shift in focus started by IACP away from materials and processes toward industry. The Jackson's Mill group also expanded the curriculum base to: (1) add communication and transportation to construction and manufacturing as components of industry; (2) study both industry and technology; (3) study industry and technology within the contexts of the universal system model, the past, present and future, a global perspective, and the social and natural environments; and (4) consider the impact of technology (Snyder & Hales, 1981, pp. 17–22).

The Jackson's Mill participants used the universal system model (input, process, output, feedback), productive processes, and managed systems models to describe the content for communication, construction, manufacturing, and transportation. These models were instrumental in directing the work of the *Industry and Technology Education Project* (Wright & Sterry, 1983) that provided examples of course structures and suggested initiation and implementation strategies. Some curriculum efforts (Indiana Industrial Technology Education Curriculum, 1992; Wood, 1987) used the models to guide much of their curriculum development efforts.

A Conceptual Framework for Technology Education (Savage & Sterry, 1990). In the nine years that followed the Jackson's Mill effort, curriculum developers continued to search for more relevant curriculum models, better ways to identify and organize learning outcomes that respond to the changing needs of society, and effective and relevant instructional strategies that help young people learn the essentials needed to manage a rapidly expanding knowledge base. The earlier subject-oriented curriculum innovations are now challenged by a process orientation in which students are expected to learn thinking skills such as problem solving and decision making. The Technological Method Model (Savage & Sterry, 1990, p. 20) describes the "Conceptual Framework's" central process. In this model, problem-solving techniques employ technological processes and resources to convert challenges into accomplishments that satisfy human wants and needs and produce consequences.

In this model, construction technology began to lose its identity. The most detrimental effect on the construction program occurred when manufacturing and construction were joined to form production. Perhaps, because of the higher cost for construction materials and the need for more space to conduct construction activities, manufacturing content is most often selected as the example to communicate material processing and management technology in the production system. Sometimes the connections and differences between the two systems are pointed out, but often it is only a verbal comment or it is ignored.

The Technological Actions Approach (Wright, 1993, pp. 31–34). Because of a perceived lack of agreement on the principal focus and direction for technology education, R. T. Wright described four "technological actions" (designing, producing, using, and assessing) in four "technological contexts" (communications, construction, manufacturing, and transportation) in two "societal contexts" (personal and commercial or industrial) that are used to produce artifacts and systems to satisfy wants and needs by solving problems and responding to opportunities. This model is useful to curriculum developers because it identifies the four basic and significant actions effective citizens of a technological world employ and structures them in terms of their unique sequential actions.

A CONCEPTUAL MODEL FOR THE STUDY OF CONSTRUCTION

A major effort to systematically identify and validate a taxonomy of subject matter for construction was completed by the IACP in the mid-sixties. New books and revisions of earlier books are available to technology teachers for teaching construction technology at the high school level (Fales, 1991; Henak, 1993; Horton, et al., 1991; Huth, 1989, Landers, 1983). Since that significant work, efforts (Henak, 1991, 1993) have been focused on adjusting the original work: (1) to present the project delivery process (PDP) described in *The Project* (Haviland, 1987), which is Volume 2 of the *Architect's Handbook of Professional Practice* and as the "process for project delivery" detailed in *Quality in the Constructed Project* (American Society of Civil Engineers, 1990, xxi-xxiv), and (2) to incorporate concepts such as fast-tracking, design-build, total quality management, and partnering into more flexible, problem-centered and student-directed courses.

Project Delivery Process (PDP)

The PDP is the core of a subject-oriented construction program. It begins with an owner's needs and ends with the owner using and assessing the structure. Figure 1–4 is a conceptual model that shows the three process organizers and how they relate to each other. Those who design structures start with an owner's needs, then create solutions, and use written and visual documentation to communicate their solutions to decision makers and constructors. Constructors use the documentation provided by designers to guide the contracting, building, and transferring of the structure's ownership. Owners use structures by occupying them and, when necessary, altering them to satisfy their changing needs.

The project delivery process (PDP) describes how people convert their needs or wants into structures. The process has six phases—predesign, schematic design, design development, construction documents, construct, and occupy/use. At each step, progress is documented with written and graphic documents and a form of pricing is used. Figure 1–5 shows how the phases, tasks, and context of the PDP relate.

PHASE I - Predesign.   Predesign services define the project. Answers are sought to questions regarding: What . . .?, When . . .?, How much . . .?, and Where . . .? Programming is a vital step in the predesign phase and is essential before any real progress can be made. It serves three purposes. First, it is an orientation time in which the architect/engineer (A/E) learns about the owner, the project, and each other's role, authority and responsibility. A second purpose of a program is to direct the design process. A

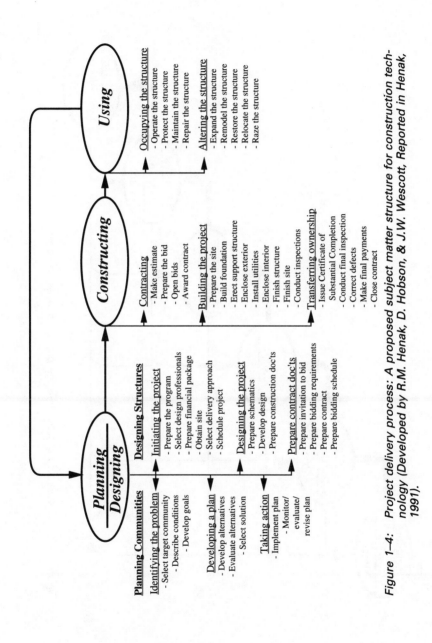

Figure 1–4: Project delivery process: A proposed subject matter structure for construction technology (Developed by R.M. Henak, D. Hobson, & J.W. Wescott, Reported in Henak,
1991).

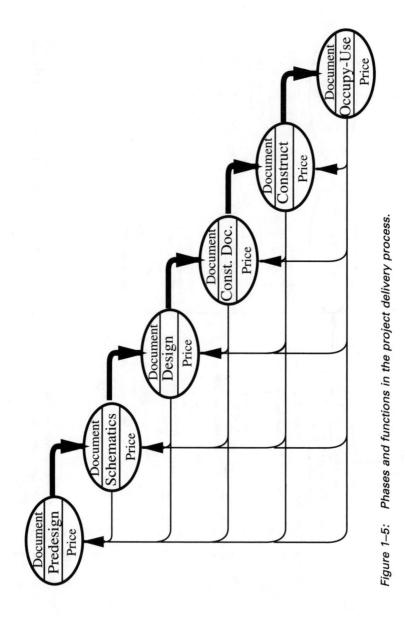

Figure 1–5: Phases and functions in the project delivery process.

finished program is a written and graphic document that presents background details and judgments useful in designing the project. A third purpose is to secure support. The financial package is designed to convince investors that the project is worthy and that there is a return on the investment.

PHASE II - Schematic Design. Making schematics is the first step in finding a solution to a design problem. The designer makes simple sketches of alternative designs for space and functions. Designers prepare sketches that address the project's program requirements and the basic concept. The sketches are little more than maps of the problem.

PHASE III - Design Development. The heart of the PDP is design development. More concern is now given to the details and less to the broad concept. The purpose is to fix and describe the size, form, and image of the entire project.

PHASE IV - Construction Documents. After the design development documents are approved, the team is allowed to prepare construction documents. These documents include working drawings, specifications, and the bidding package. Working drawings describe the physical details of the project such as the size, shape, and placement of each part of the structure. Specifications (often called "Specs") are written instructions about the project that are easier to put into words than into drawings.

PHASE V - Construction. The most help to construction architects and engineers (A/Es) can give is to produce clear error-free documents. Prior to and during construction, A/Es serve as agents for the owner who act in the owner's behalf for pay. During construction, A/Es administer the construction contract. A/Es help select the contractor and decide if the work conforms to the construction documents. Some A/E firms offer construction management services. Construction managers obtain and schedule people, materials, machines, and money to build the project as it was designed.

PHASE VI - Occupy/Use. Owners learn to use and maintain the structure and plan for the future. These services fall into four groups — initial occupancy, a post-construction review, first year warranty, and long-range planning to update the structure.

Functions

At each phase in the design of the project, accurate pricing and careful documentation is provided.

Pricing. Pricing is started early in the project and is based on a stable framework in order to meet the owner's budget. Many owners put in the contract that designs must meet their budgets.

Documentation. Written and graphic methods are used to record the ideas, processes, and decisions used throughout the PDP. Most are kept or used in making presentations to the owner or to regulatory agencies in efforts to receive approvals.

Project Delivery Approaches (PDAs)

Some owners prefer to design and build projects with their own people. Large projects may consist of one, two, or many prime contracts with the owner. Separate contracts may be awarded to clear the site, design the structure, build it, and design and build supporting roads and parking. Over the years several project delivery approaches (PDAs) have evolved, but there are three that are the most common ones. All others are variations brought about by special cases.

Linear. In using the linear method, each phase is finished before the next phase is started. The A/E determines the owner's needs and wants, creates a design, secures approval, draws up all required documents, and advertises for bids or negotiates the price. Upon the award of a contract, construction begins and when completed, the structure is put into use. This method is shown in Figure 1–6.

Fast-track. The fast-track method is faster because the design and construction phases overlap. The intent is to shorten the time needed to deliver a project and reduce construction costs due to inflation, especially when building large projects during times of high inflation. See Figure 1–7.

Design/Build. When the design/build approach is used, the client receives a firm price for creating and constructing a project. Figure 1–8 illustrates this approach. Since the price is given at an early stage, the project program and criteria are done early and are not changed much during the project's design and construction. A second advantage is that there is only one prime contract with the owner and only one point of responsibility. Other approaches have at least two points of responsibility—the designer and builder.

The Players

We all participate at some level in the construction industry. Everyone either lives *in* or lives *with* structures. However, the key players are owners, design professionals, constructors, related businesses, regulators, and users. Figure 1–9 shows the key players in the American building enterprise.

Owners. About one third of construction business in America is awarded by public owners. Public projects make up the infrastructure which includes streets, water systems, sewers, public buildings, schools, and parks.

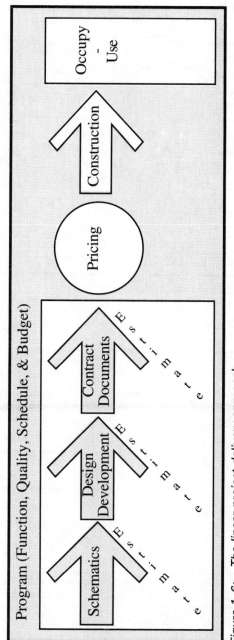

Figure 1–6: *The linear project delivery approach.*

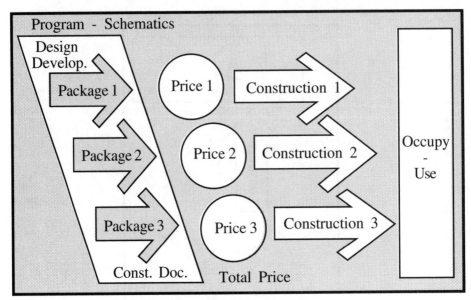

Figure 1–7: The fast-track project delivery approach.

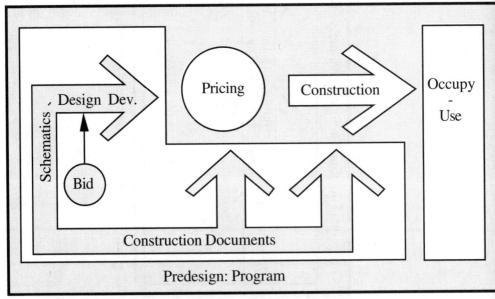

Figure 1–8: The design-build project delivery approach.

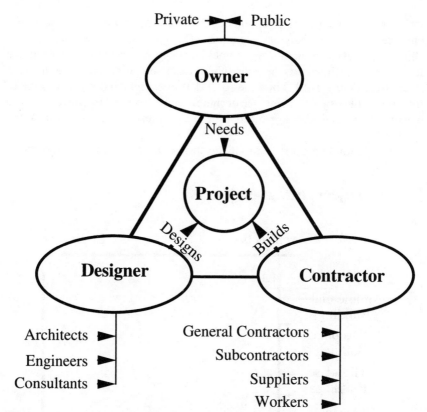

Figure 1–9: People who need, design, and build the structure play key roles.

Much of the construction on the infrastructure is repair or replacement work.

The other two-thirds of the nation's construction is started by private owners. Individuals or businesses may build once in a lifetime, in contrast with large firms that may have one or more projects in progress most of the time.

Design Professionals.   Design professionals are paid to find solutions that respond to the owner's needs and wants, meet the codes and standards, and fit the setting. They are known as planners, architects, engineers, and consultants. Planners are concerned with large scale projects that involve the formation and management policy and ranges from a neighborhood to a region or nation. Architects are qualified and licensed design professionals who design one or a few buildings, and create, coordinate, and communicate the project's overall concept and all of its parts.

There are two groups of engineers in the construction industry—design engineers and construction engineers. Design engineers work with architects to design foundations, frames, mechanical systems, and drainage for buildings. Engineers are the chief designers for heavy engineering structures. Figure 1–10 shows the level that engineers and architects relate to a range of projects. Principal designers have a prime contract with the owner and consultants are hired by the designers to provide contracted services.

Construction engineers see that the project is efficiently built the way it

Projects designed by . . .

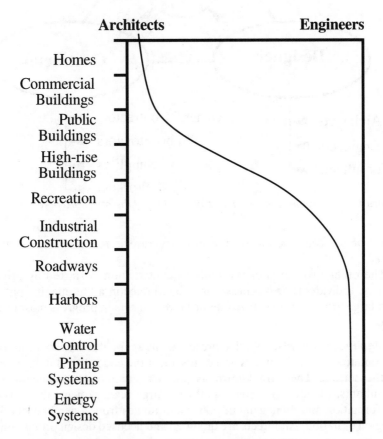

Figure 1–10: *Architects and engineers specialize in different kinds of projects.*

was designed. They work at the site throughout the building process and make sure the prescribed materials, processes, and workmanship are used.

Constructors. Constructors produce the structures envisioned by the designers. They include general contractors, developers, specialty contractors, suppliers, and labor. General contractors (often shortened to "general") work under a signed agreement (contract) with the owner to build the project. A general provides goods, services, labor, materials, and management skills. In return, the firm is compensated with a set amount of money by the owner.

Contractors build for someone else; developers build for themselves. They coordinate both the construction and marketing of the structure. Housing tracts, apartment complexes, and shopping centers are sometimes built this way. Specialty contractors, serving as subcontractors, often, do most of the actual work on a project. Each one completes a specific portion of the work such as earthwork, footings and foundations, framing, concrete, electrical, plumbing, roofing, glazing, and hanging drywall.

Suppliers provide materials, products, and equipment. Material and product suppliers provide and sometimes install components that become a part of the structure. Construction equipment suppliers provide highly specialized equipment used by contractors to move or install items, provide brief support, and heat or clean the site. The physical work of building structures is done by skilled and unskilled labor.

Related Businesses. Without the work of real estate, finance, insurance, testing, and legal professionals, construction projects would not be realized.

Context Of Construction

The context in which construction functions is characterized by increasing complexity, limiting regulations, intensifying risk, expanding teamwork, and growing emphasis on quality.

Building Is Complex. Structures are becoming more and more involved because new materials and processes are being developed each day. We are also learning how to build larger structures and to integrate more diverse uses and complex mechanical systems within single structures.

Regulation Is Severe. In an attempt to prevent negative impacts, codes and regulations governing the design and construction of structures are becoming more precise. In the past, one could reflect on past actions to plan future actions, but this is no longer a good practice. Many new structures are built to last a long time and are so large that their impact on the environment, people, and the economy is far greater, and the impact will last longer than those of the past.

Risks Are High.   Large sums of money, reputations, resources, and satisfaction are on the line when one builds. Mistakes in structures are hard to hide and expensive to correct.

Team Oriented.   Employers repeatedly stress the importance of being able to work with a team on common objectives. Project teams are used to design and build structures because construction has become so complex and so much is at stake. Project teams usually include representatives of owners, design professionals, and constructors who work to effectively control costs, time, and quality. Figure 1–11 illustrates this concept. Supporting roles are played by related businesses, regulators, and users of the completed structures. A new trend, partnering, has been developed and promoted by the AGC of A (Jackson, 1992b). It is an alternative way to resolve disputes that often lead to costly, time-consuming formal lawsuits.

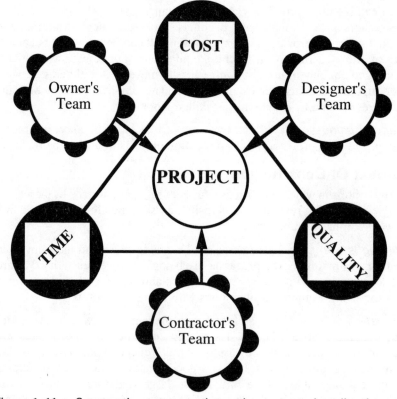

Figure 1–11:   Construction teams work together to control quality, time, and costs.

Quality Is Emphasized. Quality in a constructed project is defined as meeting the requirements (quality, schedule, and budget) of the owner, design professional, and constructor, while conforming to regulatory rules, laws, standards, and other matters of public policy (American Society of Civil Engineers, 1988, pp. 1–2). The AGC of A has addressed improving quality by adapting the total quality management (TQM) concept to the construction industry. Bill Harrison (1992), the president of the Phoenix Leadership Institute that specializes in TQM, stated "Total Quality Management is about solving problems permanently" (p. 30). Everyone, not just management, is involved. Harrison adds, "The key to success with TQM is training and empowering all members of the team to make quality improvements, adjustments, and decisions on their own" (p. 30).

A CONSTRUCTION TECHNOLOGY CURRICULUM FOR THE FUTURE

A comprehensive program should include significant experiences in designing and engineering, constructing, and planning the built environment; engage students in meaningful, creative problem-solving situations related to their lives; address current concerns and those in the future; and communicate an accurate view of the opportunities available in the construction industry.

Designing And Engineering

Activities in designing and engineering the built environment center on the work of the design professionals — architects, engineers, and design consultants — and their interaction with public and private owners, constructors, related businesses, and regulatory agencies. The scope of the subject matter begins with the initiation of the project and ends with completed contract documents. A course entitled Construction Planning and Design (Indiana Industrial Technology Curriculum, 1992c) reflects the intent of a construction design experience.

Constructing. The principal concern of an experience in constructing the built environment is to provide activities that include contracting, building, and transferring ownership. Projects that incorporate as many building processes as possible are used for the major activity. Scheduling, managing cash flow, initiating change orders, performing periodic inspections, and making partial payments should be integrated into the project. In this course, the constructor team plays the lead role. Consideration should be given to the potentially adversarial relationship between designers, constructors, and owners as they try to manage quality, time, and cost. Constructing Structures (Indiana Industrial Technology Curriculum, 1992b) is a course designed for this area of study.

Planning. Planning is a fascinating area of professional practice. Although there are many exceptions, architecture is primarily concerned with the physical design of individual buildings, planners are concerned with many buildings or entire regions. The planning process includes three phases — description, prediction, and action. Figure 1–12 illustrates the process and the breadth of the subject matter. Activities are designed to provide plans of action for resolving current, community concerns. Students may write a comprehensive plan and present it to city or county officials. Community Planning (Indiana Industrial Technology Curriculum, 1992a), provides a model for this experience.

CONCLUSIONS

This chapter challenges technology educators to create construction courses that produce a populace that is literate about construction, is able to participate in addressing current problems and issues, and will reduce the potential of having a critical labor shortage in the near future. It suggests that broader goals and subject matter should be identified and that more engaging strategies and realistic settings should be used. Three courses were described in which the PDP provided the subject matter organizers. Key personnel was used to focus the activities. In each course, students: (1) are introduced to a range of construction technology concepts, (2) are expected to apply the concepts to realistic issues and problems, and (3) play authentic career roles.

There is a great need for an activity-oriented construction curriculum that is organized around using problem-solving skills and resolving real-life problems and issues, while fostering basic construction literacy. Activities should demonstrate the relevancy of construction to important aspects of the students' lives, develop their problem-solving, critical-thinking and decision-making skills, and foster positive attitudes. Well-designed learning experiences can provide young people with sufficient understanding to appreciate the construction industry and its contributions, help them participate effectively in a democracy, inspire capable students to enter construction careers, and provide the populace with sufficient understanding to appreciate things technologically. Construction concepts must be taught in ways that help *all* students develop better understandings of their world.

ELEMENTS OF A PHYSICAL PLAN	Environmental Protection	Land Use	Transportation	Housing	Economic Development	Public Facilities-Utilities/Services	Fiscal Plan	Historic Preservation
DEVELOPMENTAL PHASES								
DESCRIPTION PHASE: Diagnose Problem Set Goals								
PREDICTION PHASE: Develop Alternatives Evaluate Alternatives Select Solution								
ACTION PHASE: Implement Plan Monitor/Evaluate/ Revise Plan								

Figure 1–12: Subject matter model for community planning (Indiana Industrial Technology Education Curriculum, 1992).

REFERENCES

American Society of Civil Engineers. (1990). *Quality in a constructed project.* New York: American Society of Civil Engineers.

Betts, M.R., Fannin, J.W., & Hauenstein, A.D. (1976). *Exploring the construction industry.* Bloomington, IL: McKnight.

Brady, M. (1989). *What's worth teaching?* Albany, NY: State University of New York Press.

Cochran L.H. (1968). A comparison of selected contemporary programs in industrial education. *Dissertation Abstracts International, 30,* 3A. (University Microfilms No. 69–14, 664)

Cochran, L.H. (1970). *Innovative programs in industrial education.* Bloomington, IL: McKnight & McKnight.

Commission on Reform of Secondary Education. (1918). *The reform of secondary education: A report to the public and the profession.* New York: McGraw-Hill.

Dyrenfurth, M.J. (1991). Technological literacy: A new paradigm. In M.J. Dyrenfurth & M.R. Kozak (Eds.), *Technological literacy* (p. 7). Mission Hills, CA: Glencoe.

Dyrenfurth, M.J., & Kozak, M.R. (Eds.). (1991). *Technological literacy.* Mission Hills, CA: Glencoe.

Educational Policies Commission. (1961). *Central purpose of American education.* Washington, DC: National Education Association.

Fales, J.F. (1991). *Construction technology: Today and tomorrow.* Mission Hills, CA: Glencoe.

Fink, D.L. (1984). *The first year of college teaching.* San Francisco, CA: Jossey-Bass.

Gasperow, R.M. (1988). The future need for skilled workers. *Constructor, 70*(10), 12–14.

Glatthorn, A.A. (1987). *Curriculum renewal.* Alexandria, VA: Association for Supervision and Curriculum Development.

Harrison, B. (1992). TQM and the small contractor: The time to adopt is now. *Constructor, 64*(9), 30.

Haviland, D. (Ed.). (1987). *The project (Volume 2)*. In *The architect's handbook of professional practice*. Washington, DC: The American Institute of Architects.

Hawkes, N. (1990). *Structures: The way things are built*. New York: Macmillan.

Healy, W.K. (1988). The skilled worker crisis. *Constructor, 70*(10), 15–16.

Henak, R.M. (1991). Construction education for the 21st century. *School Shop/Tech Directions, 50*(7), 9–11.

Henak, R.M. (1992). Effective teaching: Addressing learning styles. *The Technology Teacher, 52*(2), 23–28.

Henak, R.M. (1993a). *Exploring construction*. South Holland, IL: Goodheart-Willcox.

Henak, R.M. (1993b). *Exploring construction-Instructor's manual*. South Holland, IL: Goodheart-Willcox.

Henak, R M. (1993c). *Exploring construction-Student activity manual*. South Holland, IL: Goodheart-Willcox.

Henry, W.R. (1989). AGC President Paul Emerick. *Constructor, 61*(4), 16–19.

Hill, J.R. (1981). *Measurement and evaluation in the classroom*. (2nd ed.). Columbus, OH: Charles Merrill.

Horton, A., Komacek, S.A., Thompson, B.W., & Wright, P.H. (1991). *Exploring construction systems*. Worcester, MS: Davis.

Huth, M.W. (1989). *Construction technology*. (2nd ed.) Albany, NY: Delmar.

Indiana Department of Education. (n.d.). *Technology preparation curriculum*. [Brochure]. Center for School Improvement and Performance, Room 229, State House, Indianapolis, IN.

Indiana Industrial Technology Education Curriculum. (1992). *Community planning*. Muncie, IN: Center for Implementing Technology Education.

———. (1992). *Constructing structures*. Muncie, IN: Center for Implementing Technology Education.

———. (1992). *Construction planning and design*. Muncie, IN: Center for Implementing Technology Education.

———. (1992). *Introduction to construction*. Muncie, IN: Center for Implementing Technology Education.

Jackson, R.H. (1992a). Construction's role in economic recovery. *Constructor, 74*(11), 5.

Jackson, R.H. (1992b). Partnering for success. *Constructor, 74*(11), pp. 45–46.

Kozak, M.R., & Robb, J. (1991). Education about technology. In M.J. Dyrenfurth & M.R. Kozak (Eds.), *Technological literacy* (pp. 28–50). Mission Hills, CA: Glencoe.

Landers, J. (1983). *Construction: Materials, methods, careers.* South Holland, IL: Goodheart-Willcox.

Lux, D.G., & Ray, W.E. (1970). *World of construction.* Bloomington, IL: McKnight.

Lux, D.G., & Ray, W.E. (1970a). *World of construction-Laboratory manual.* Bloomington, IL: McKnight.

Lux, D.G., & Ray, W.E. (1970b). *World of construction-Teacher's guide.* Bloomington, IL: McKnight.

Lux, D.G., Ray, W.E., Blankenbaker, E.K., & Umstattd, W. (1982). *World of construction.* Bloomington, IL: McKnight.

Marzano, R.J. (1992). *A different kind of classroom: Teaching with dimensions of learning.* Alexandria, VA: Association for Supervision and Curriculum Development.

National Commission on the Reform of Secondary Education. (1918, 1973). *The reform of secondary education: A report to the public and the profession.* New York: McGraw-Hill.

Pollio, H.R., & Humphreys, W.L. (1988). Grading students. In J.H. McMillan (Ed.), *Assessing students' learning.* San Francisco, CA: Jossey-Bass [New directions for teaching and learning series (No. 34)].

Pruit, D.J. (1991). *Partnering: A concept for success.* Washington, DC: Associated General Contractors of America.

Savage, E.N., & Sterry, L. (1990). *A conceptual framework for technology education.* Reston, VA: International Technology Education Association.

Secretary's Commission Achieving Necessary Skills. (1992). *Learning a living: A blueprint for high performance.* Washington, DC: U.S. Department of Labor.

Short, J. (1991). Pre-construction teamwork. *Constructor, 73*(6), 30–32.

Snyder, J.F., & Hales J.A. (1981). *Jackson's Mill industrial arts curriculum theory.* Charleston, WV: West Virginia Department of Education.

Sullivan, M.R. (1989). Construction ahead: The goals of the CIWF. *Constructor, 71*(2), 18–20.

Towers, E.R., Lux, D.G., & Ray, W.E. (1966). *A rationale and structure for industrial arts subject matter.* Columbus, OH: The Ohio State University.

Tyler, R.W. (1975). *Basic principles of curriculum and instruction.* Chicago, IL: The University of Chicago Press.

Wood, J. (1987). *Exploring construction technology.* Stillwater, OK: Mid-America Vocational Curriculum Consortium.

Wright, R.T., & Henak, R. M. (1993). *Exploring Production.* South Holland, IL: Goodheart-Willcox.

Wright, R.T., & Sterry, L. (1983). *Industry & technology education: A guide for curriculum designers, implementors, and teachers.* Lansing, IL: Technical Foundation of America.

Wright, R.T. (1993). Lack of agreement on principal focus and direction for technology education. *Camelback symposium: A compilation of papers.* Reston, VA: International Technology Education Association.

Zemke, R., & Zemke, S. (1981). Thirty things we know for sure about adult learning. *Training, 18*(6), 45–52.

CHAPTER 2

Past, Present, And Future Of Construction Technology

David A. Ross, Ed.D., Department of Industrial Technology
Georgia Southern University, Statesboro, GA

CONSTRUCTION IN THE PAST

It is probably safe to say that mankind has always possessed the need to build. Unlike other areas of historical significance, construction lacks documentation of important names, dates, and locations. The first time someone dug a hole to secure a post or laid stones to form a wall were significant events in construction. The first builders were probably less bewildered and confused as to what to build and where to build it, than those who design, build, and use contemporary construction projects. Early builders undertook projects in order to meet a few easily defined needs such as shelter and storage; but the contemporary constructor fabricates structures for a plethora of purposes which often depend on complex technologies.

Early Constructors

We do not know why the first structure was built, who built it, what materials were used, or how it was built; but it seems that there has always been a compelling instinct for people to continually improve their living conditions by constructing structures. The earliest cultures knew how to force posts into the ground to support platforms. As people became more advanced, other natural materials such as mud was used to secure the posts.

The first constructors built simple shelters, but materials and structures failed due to natural elements and war (Ewald & Williams, 1970). But constructors are survivors; nothing seems to stop their will to build. As population grew and the demand for more and better structures increased, new methods, tools, and materials were developed; crafters became more skilled and innovative; and buildings, cities, and communities were built.

Early construction projects were fairly simple. Where stone was not available, walls were made by laying sun-baked bricks in successive courses. Whitaker (1936) stated, "The first dated examples of brick construction was located in Egypt" (p. 18). Brickmakers used water, sun, and clay to make their bricks; and only after years of practice, could bricklayers form walls and roofs of structures that would not collapse.

Early Structures

Camesasca (1921) envisioned that the first structure may have been a round cone or tent-like shape made of natural materials; or perhaps, it was a square or rectangular shaped structure of thatch and mud held vertically by stakes driven into the ground. Regardless of how the first structure was designed, it did not take long for the four-walled square-angled building to take prominence as the norm. By joining the tops of the walls with flat bundles of material or by having one wall higher than the other, a slope was produced to shed water and protect the structure and interior space. The more severe the weather, the steeper slopes became. Later, designs changed to include a point or ridged assembly on top to produce the gabled roofs we use today. Openings were added for access, light, and ventilation. When it was discovered that space could be expanded by adding walls, roofs, and wall openings, the space elements of structures were formed.

A Change In Direction

A significant shift in thinking from providing secure and stable structures for life on earth to concerns about life after death influenced construction greatly. For thousands of years, Egyptian tombs, temples, and palaces were on the minds of kings and priests; and thus, dominated the work of builders rather than the needs of the total population. According to Whitaker (1936), the perceived urgency to build for protecting and glorifying life after death caused constructors to master limestone, sandstone and granite and to develop a science of construction.

An enormous heritage of building design was derived during the age of Greece. The Greek constructors preferred to use their labor to advance the comforts of living. Rather than use their time and resources to build awe-inspiring temples, they invented rooms with flush toilets and running water. Unfortunately, little is left of the houses; and accounts of that era only reflect the building of temples, theaters, and a few monuments.

Origins Of Communal Planning

As population increased, there was a shift of focus from providing shelter to concerns about the general arrangement of buildings and to the planning process. Planning dealt, not only with the single structure, but also with how structures related to each other, to culture, and to life. The skill of constructors was used for building towns for the purpose of providing comfort and convenience rather than single structures. Thus, concerns for private shelter evolved into communal planning.

As a communal economy evolved around a close, simple, and direct lifestyle based primarily on food, constructors gained a clearer direction on what and where to build. The early builders' responsibility was to make life progressively easier and more agreeable for the productive people in the community. Greek constructors were concerned with unity and harmony among the parts of a structure and their success was determined by the care they took in planning. Skilled constructors used their talents to build comfortable and convenient towns rather than single structures.

Roman Contributions

The Romans felt that the building of an empire had to show constant progress and used their structures as symbols to promote the Empire. Decorations and embroidery work were used to embellish huge theaters, arenas, baths, arches, and structures which were built to amuse and impress the populace. Since speed was of utmost importance, the masonry arch contributed greatly to achieve their goal and the concrete arch even more so. Romans learned to pour concrete arches and vaults in one piece and required far fewer people than when building masonry arches. The use of concrete was the first major labor-saving innovation since the discovery of the post-and-lintel technique.

When the Roman empire finally fell, many building techniques remained and are still used. Specifically, the Roman constructor discovered how to make stone walls with less labor, build tight fire-proof roofs, use stucco-concrete mosaic, and replace post-and-lintel construction with a composite. If they had planned and built a civilization rather than an empire, they may have contributed even more.

Immigrant Constructors In Early America

The first structures in America were built by native Americans. Cliff dwellings in the west and burial grounds found throughout the continent are examples. The first immigrants found an abundance of land and forest, and virtually no structures. People by the thousands poured into this unspoiled land and filled the silence of the forest with the noise and clutter of construction. Constructors from London, Bristol, Amsterdam, Stockholm, and Plymouth were the first to build in America. The structures they and their children built dominated the landscape in the 1600s. Since wood was plentiful, the initial task was to clear the land and use the wood to fabricate most structures.

The strongest demand for construction came from American constructors and developers who felt that a way to profit was to increase the population with immigrants. Immigrants came and increased the demand for land and construction of all kinds. Two economic concepts drove the construction effort in America beyond meeting basic needs. They were to profit by: (1) creating a demand for goods and (2) increasing land values.

Creating a Demand for Goods. The first construction planning in America was not done by constructors. Structures were designed to satisfy the desires and needs of English land owners and industrialists. The idea was based on exploiting the land to provide employment for English vagrants and criminals who were then able to buy goods from England.

Increasing Land Values. Social and economic factors were also involved in raising the price of land, which led to large profits for the landowners and developers. Developers knew that land owners needed workers and that people could add value to the land needed for building purposes. With the immigrants came an increased demand for land; and each time a new structure was completed the price for the adjoining land increased. They started the pattern of building high and fast and used debt to build what they could not afford.

From Crafters To Labors

For centuries, buildings were built by crafters. Kerisel (1987) characterized construction in America by the early settlers. He stated:

- There was a fondness for the use of wood in the New England States.

- Stone was used in the mid-west/Pennsylvania area.

- Bricks were a major material in the structures of the south.

- Two styles were evident — Colonial and Georgian. (p. 13)

Builders of that time could not put buildings together with machines. The crafter was still the hub of the project.

Just as America started to grow, simple buildings seemed to vanish. The feel and touch from the use of tools and materials passed out of the hands of the true crafter and fell into the group known as labor. Workers, under the direction of the contractor, followed the specifications of the designer.

Because of this change, workers no longer had input into the appearance of a structure.

The demand for larger and for more dramatic buildings that would impress others prompted the need for intricate work and, ultimately, workers who needed training. The final separation of crafter and contractor came when building and design schools were formed to satisfy the need for skilled crafters.

Urbanization

It was not until the 17th century that constructors could build as they pleased. The New England village, the missions of California, the trading post, and religious groups have all tried to balance social, agricultural, and industrial life. There was great variety in building design because clients who could pay for a builder's services did not want the same kind of structure. Each client seemed to want buildings that were longer, wider, or higher; or that had more windows, a larger stairway, or a bigger fireplace than anyone else. It was, and still is, an endless search for the best size or arrangement of parts in a search for refinement in existing structures. There were plenty of land, workers, materials, and financial supporters to promote the building initiatives; no one thought of congestion.

A significant innovation in construction was the use of steel as support members for structures. As constructors designed and erected steel skeletons, larger and taller buildings could be built and congestion became an issue. As the cities, sidewalks, and streets filled, the demand for more traffic routes became a major concern. Traffic and traffic routes fed off of each other and the development of rapid transit offered a remedy for congestion by moving people to the edge of towns. Developers had created their first inflated land prices which caused a shift in building styles. While price and debt increased, the structure's size, number and size of rooms, quality of materials, and level of decoration decreased.

CONSTRUCTION IN THE PRESENT

The present construction industry is different from the manufacturing or service industries. It is a broad mixture of skills, resources, and control groups working together around a specific project. The construction industry has characteristics that are shared with other industries: but four characteristics unique to the construction industry are: (1) the physical nature of the service; (2) the structure of the industry together with the organization of the construction process; (3) pressures of demand; and (4) pricing processes. It is important to note, that this situation is slowly changing and will continue to change; and a more conventional industrial structure is emerging.

Management

Today's construction industry can be generally described as a closely controlled, family-oriented business, which is very competitive. Innovative methods and financial data are kept as top secrets and most contractors apply management principles as an antithetical process which makes it difficult to provide a sound statistical base for analysis and projection. It is only possible to make generalized statements and inferences about the current status of the construction industry.

The construction industry is adapting management principles used by conventional manufacturing industries, such as Just-In-Time (JIT) and Total Quality Management (TQM). Barrier (1992) stated that, "TQM has become the most popular abbreviation since TGIF. It is an all-encompassing term that stands for the quality programs that have spread through most of the U.S. businesses" (p. 86). The TQM concept was initially adopted by large manufacturing companies that faced aggressive challenges from Japanese competitors. Diverse organizations outside of manufacturing such as education and construction are adopting its principles and are improving the quality of their services and structures. Barrier (1992) listed four essential ingredients of a TQM program.

1. An intense focus on customer satisfaction.

2. Accurate measurement, using a fistful of readily available statistical techniques of every critical variable in the company's operation.

3. Continuous improvement of the products and services.

4. Most importantly, new work relations based on trust and teamwork (p. 90).

Structure Of The Construction Industry

The heart of the construction industry consists of construction contractors, subcontractors, investment builders, and design/engineering companies. See Figure 2–1.

The percentage of people that make up each group is illustrated in Figure 2–2.

The largest organization from this group are the design/build companies ranging from giant industrial corporations with contracts in the

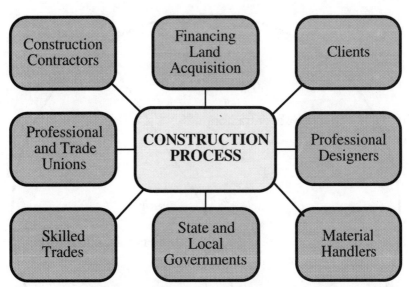

Figure 2–1: *People who make up the structure of the construction indus-try (Dodge & Sweet, 1992, p. 28).*

billions of dollars, to small operators who hire staff and subcontract the work.

Contractors are classified as general contractors and specialty contractors (Dodge & Sweet, 1992, p. 14). General contractors coordinate jobs, provide supervisory services, and, normally, perform only part of the actual work on the project. Specialty contractors complete limited aspects of the construction work and are referred to as subcontractors. Specialty contractors are sometimes hired directly by a client to do a particular type of work such as maintenance and alteration work.

In most cases, employees working for a contractor are not permanent employees. For example, Lefkoe (1980) stated:

> One half of the nation's residential construction is built by non-union workers and a small amount of non-residential construction is built by non-union workers. With the exception of residential, the bulk of the new construction is being built by union members. The majority of these craft unions are affiliated with the Building and Construction Trades Department of the AFL-CIO. (p. 41)

Construction employees are assembled for specific projects and are often employed by different contractors during the course of a year. When a project is finished, employment is terminated unless the employer has a new contract for a similar job.

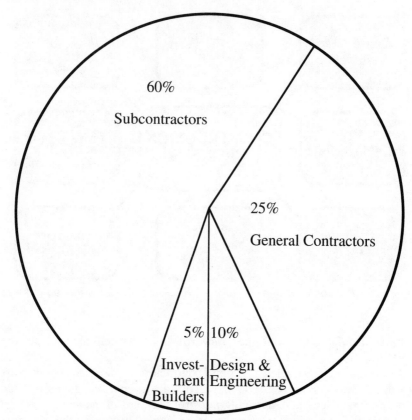

Figure 2–2: Percentage of people employed in construction by category (Dodge & Sweet, 1992, p. 44).

What Does Construction Provide?

The final products of the construction industry are expensive and large in scope. They require considerable land, are customized to the requirements of individual customers, and involve a variety of materials and components; and many of the components are manufactured somewhere else. Hillebrandts (1985, p. 29) identified four ultimate uses of constructed structures. They are: (1) to further production (e.g., factory/industrial buildings); (2) to improve and expand the infrastructure of the economy, (e.g., roads); (3) to meet social requirements; (e.g., hospitals); and (4) to satisfy personal needs and enjoyment (e.g., housing).

McNiel (1992) reported that the construction industry is the largest industry in the U.S.; and those that participate must be accountable for the effect they have on the economy. The industry organizes, moves, and assembles materials and component parts to produce durable buildings, public works, and structures. Because of its size, the economic importance of the construction industry is often used as an indicator of the economy. A change in the level of the construction industry's output affects employment, incomes, demand, and ultimately the output of other sectors of the economy. If there is a decline in the level of construction activity, a downward spiral effect is felt in other industries and on the economy as a whole. Demand for its products and services is also related to the rate of growth and consumer demand because the industry is an investment industry.

Construction Is Risky

Risks are high in the construction industry but so are the potential rewards. Homebuilding, which is the most volatile component of the construction industry, is presently a strong force in the overall construction industry showing a steady growth and an increase of new companies over the last ten years.

The structure of the current construction industry and its method of operation is perhaps the single most important factor responsible for the uncertainty of the industry. The construction industry is not only risky; it is also highly competitive. Construction companies who focus on the private market tend to be more lucrative than those depending on public contracts. Public projects require large investments in equipment required for heavy construction; and public contracts are often more competitive.

The industry lacks the ability to control its own market. What contractors produce is primarily controlled by individual clients and their wishes. Combined opportunism, intuition, market experience, and rough estimates of future demand guide their clients' requests. Many contractors find experimentation risky as it relates to long-term planning; and most construction companies have little or no cushion for failure.

While companies do operate under high risk, most of the risk and failure is associated with conventional construction methods in conventional ways for traditional structures. Kerisel (1987) lists six fundamental reasons for the construction industry's conventional approach:

1. the industry's structure and its method of operation,

2. seasonality in construction,

3. building codes,

4. licensing of contractors and certain craftspersons,

5. union power, and

6. the concept of jurisdiction (p. 33).

The construction industry is subject to fluctuations in demand that should be studied and monitored. Information systems that increase the knowledge about changing demand and about forecasting are potential ways to reduce undesirable effects of these fluctuations.

CONSTRUCTION IN THE FUTURE

The following objectives were used to describe the construction industry's future:

1. *Trends* - Practices that will have the greatest impact upon the culture, society, and environment and upon the processes and people who participate in change.

2. *Challenges* - Problems which the changes from these influences would present to the construction industry.

3. *Projections* - Indicators that suggest and direct actions which should be taken to reach targeted objectives.

Trends

Change is inevitable and will come in response to the increased demands of an expanding population. According to Dodge & Sweet (1992), three trends will speed the rate of change in the structure and output of the construction industry. They are:

1. *Technological advances* will have a major and dramatic impact on the internal structure, management, and methods of operation of the building industry over the next few years and will have an effect on the industry's final product.

2. *The public and private sector relations* will become more coordinated. Positive, participatory groups will place importance on the building goals, planning policy issues, and ways to reduce costs.

3. *Additional pressure from market growth, shifting national priorities, and entry of organizations from outside the traditional building industry* will cause an up-sizing in the planning and building enterprises so they handle larger projects, longer planning time frames, and more attention to human need, stress, and environmental factors (pp. 213–214).

These three categories will produce a radically different construction industry in the future. Of the three mentioned, it will be the technological advances that will cause a rapid growth and major impact on the industry's structure and methods of operation. Innovations in technology will play a role in the way construction is viewed as a generator of new materials and products, new engineering concepts, new subsystem technologies such as energy, waste handling, and the movement of people and products.

Technological Advances. The rate at which the residential construction industry responds to new technologies will accelerate. It will: (1) follow the trend toward pre-manufactured items such as hardware, plumbing fixtures, and windows and doors; (2) extend to large units like kitchens and bathrooms; and (3) then, encompass entire structures. In commercial construction, current, field-fabricated systems will be replaced with manufactured systems and components for walls and ceilings that require only on-site assembly.

New trends will tend to make the hand-crafted construction systems more costly than their pre-manufactured counterparts. The result is that most residential and commercial projects will have a more manufactured look. Furthermore, the hand-crafted portion of a project will represent an even more desirable design feature and will still be in demand as a personal expression of preference by those who can afford it. Contractors competing for projects will become more innovative in selecting and using systems to reduce look-alike boredom — a characteristic of many manufactured systems.

As for the structures, they will become less differentiated as to use; and construction will be planned to permit continued change for evolving activities and functions. The larger projects, such as shopping centers and resort areas, will extend to structures intermixed with working, residential, recreation, and institutional needs.

Public and private sector relations. Street (1992, p. 20) suggests that we open dialogues between a variety of participants. A positive move in the development of the construction industry is more government-industry partnerships. The areas of research and development, design, fabrication, and assembly will be the most productive. By having groups join as partners, the industry will improve substandard neighborhoods throughout the U.S. The introduction of new ideas, techniques, and methods for problem solving will have a long-term benefit on the industry.

The present construction workforce will be regrouped into new organizations with new alliances. Many of those in the present industry with commitments and investments in the current system will reject change and be unable to cope and drop out. The voids they leave will be filled with new and inexperienced people. As the changes happen, contractors must find the means for communicating and training the present workforce in understanding the new needs of people with the objective of providing the best environments that a socially and technically advanced construction industry can provide.

Growth, Priorities, and Outside Organizations. We are driven by population growth, migration patterns, technology, public policy, and market growth (Christie, 1992, p. 15). The demand for structures will continue to outgrow the available supply. The extent of new construction developments in the next few years will be unmatched by anytime in history and, we hope, will lead to a higher standard of living and a better environment for everyone. A critical factor in the provision of state financing for the construction industry will be new government policy.

Land cost will continue to increase sharply as a percentage of total-project cost will lead to an intensification of land use. The pattern of rapid and vital suburban growth will continue to use large amounts of space, while a revitalization of urban centers increases. Because of this demand, the construction industry's volume will grow faster than the supply of trained and skilled designers, contractors, project managers, and labor.

The direction of change is likely to be toward a more conventional, large-scale, industrial structure integrating new management functions. Information systems are being created that group all functions and participants within the construction industry on a national information retrieval system. The expansion of computer technology will make the automation of information-intensive activities possible and profitable.

Challenges

In an age of accelerated social and technological change, a corresponding change in all aspects of construction can be expected. The present challenge is to understand, anticipate, and use change as an opportunity to improve efficiency and increase skills.

Society, the construction industry, and the activities and responsibilities of those that work in the construction industry are quite different from what they were only a generation ago. In this age of accelerated social and technological change, constructors must expect rapid changes in their roles and responsibilities. Constructors must understand, anticipate, and use change as an opportunity to increase their skills and effectiveness by continually responding to the technical knowledge and social structures of the day.

How will these trends challenge future constructors? According to Ross (1981), these changes will include:

- management systems for the industry;

- development of participatory groups for problem-solving strategies; and

- growth in research & development on the industry and building process (p. 13).

Management. The construction industry will continue to be a social endeavor controlled by the demands of people. Management and leadership by the constructor and their actions will have the greatest influence on the environment. The success in making changes in the construction industry will depend upon the ability of constructors to strengthen the organizations they form and increase the degree to which they develop and exert influences at critical junctions in their careers.

Distinguishable within the construction industry is the contractor whose role has a long historic precedence. Emerging from the crafts and arts as master builders and then later from nobles and gentlemen scholars, contractors have only recently been formally educated for their role in industry. Regardless of the background and knowledge, the contractor of the future will stress the need to create the highest level of environment for human habitation. In spite of the wide range of individual activities in which a contractor might be engaged within the organizations, the work of the contractor will be distinguished by their ability to manage and direct others.

Participation. We are entering a time of extreme cultural and social stress created by the need and demands from low-income groups for equality of opportunity and for greater participation in decision making. Active participants, who emerge as effective leaders to deal with these issues, will have the greatest influence on what is constructed.

Construction personnel will expand to include a greater mix of professionals in the industry. Environmentalists, and people in the social and behavioral sciences, natural sciences and engineering will move in greater numbers into construction design activities. Thomas (1989) outlined five reasons for this projection. They are:

1. an increased interest in urban and environmental planning;

2. research and professional services in areas of social engineering;

3. growing pressure from government to include social and environmental factors in buildings;

4. pressure to develop total ecology approaches to building planning; and

5. more citizen participation in the planning process for large structures (p. 69).

Involvement of many professional, government, and non-professional people will take a more formal, permanent, and important place in decision making.

Research and Development.   Constructors must recognize that they are entering a critical period for the industry. A period of social reforms, realignment, and reorganization will have a profound effect on our public and private institutions, enterprises, and environment. These changes will influence the demand for innovation in communication; and the planning, designing, and selecting delivery systems to provide technological solutions for constructed projects that will equitably enhance the quality of life for all.

The constructor's purpose for adopting new management techniques is to improve control and profits while providing high-quality service to customers. In the construction industry, evolving information systems are part of a larger trend toward more use of comprehensive and sophisticated management systems to control cost, time, quality, and profitability (Ross, 1989). Some of the specific techniques that will continue to be used for control procedures such as Critical Path Method (CPM) will continue to be utilized by the constructor. For example, the project initiation phase for large developments will include more complete analysis and decision making before the design personnel begin their work. New analytic tools to determine return-on-investment, optimum-use-patterns, and construction-phasing relative to marketing schedules will be common.

A systematic approach to management is coming to the construction industry as it has to the manufacturing and service industries, and is probably the best way to sum up the influx of new techniques which are leading to the restructuring of the construction industry. A host of jurisdictional questions relating to restructuring functional responsibilities in the building process are likely to receive closer attention in the next few years. Under a systematic approach to management, decisions and conclusions reached may well change the traditional roles of most contractors.

New management systems such as TQM, JITM, and Total Participatory Management systems, which are being imported from other industries, will have a profound effect on the structuring of the building process. The largest change will concern the roles and responsibilities played by all construction participants in the systematic step-by-step approach to conceive, plan, and build a structure.

Because of the concerns expressed by environmentalist groups toward the protection and improvement of the environment, there will be efforts to improve our parks, seashores, and other public recreation facilities. This concern will also force the construction industry to use materials and processes that are environmentally safe. The public will be expected to be participants and provide support for a better public environment. Levels of support for larger improvements will tend to receive higher priority because the money will be supplied by regional and national government agencies.

Local improvements will be more difficult because the funds affect personal tax increases (Horn, 1992, p. 49).

Recruitment and Training. Statistics from the U.S. Bureau of Census (1992) reveal a growth need of 2.1% for the construction workforce each year through 1995. This translates to an annual rate of approximately 40,000 new workers. McNiel (1992) indicated that the industry will need to attract about 200,000 newcomers annually to meet the demand for construction anticipated by the year 2000.

Projections

Demand. It is safe to say that the demand may be greater than anyone can project and that much of this demand will be built at a luxury level. Increasingly, larger proportions of the population will demand living, working, and playing in structures with high-level physical comfort and psychological satisfaction. This continuing increase in standard of living for the majority of the population will permit a personal choice of place for both living, working, and pleasure. Those who will join this increasingly larger-sized middle class will expect more technologically advanced goods and better quality apartments, houses, shopping centers, restaurants, and hotels.

Constructors. The present construction industry should not worry that there will be an insufficient demand in the future. Just the opposite, contractors must recognize that there will be stronger efforts made to increase the construction industry's efficiency by attracting groups with new areas of expertise, such as environmentalists and minority interest groups. Those that are involved in the current industry will continue to be needed, but the organizational structure will change where the industry is under new controls and new management styles. This demand for greater efficiency, economy, and speed will: (1) affect the time sequence for planning and constructing, and (2) will accelerate the development of pre-manufactured, off-the-shelf systems and complete structures. More concern for design constraints will also be required. The demand for technology will be for accessories and more sophisticated, environmental innovations in mechanical and electrical systems.

Structures. There will continue to be a need for industrial and commercial buildings and for housing, which provides a benefit from direct consumption. However, many of the structures in the future will be built for the collective enjoyment or benefit of large groups of people (e.g., schools, sports facilities, and shopping complexes). The purchase of the service or benefits gained from these structures will be from groups other than the users.

A larger number of people will be involved in service-related businesses, and the rate of growth of office-type space will be more rapid than the growth for traditional factories. The needs of industry will become more aligned with those of service businesses with expanded space for human information and decision activities. The future constructor will need to be aware that the level of conveniences and the expectations of the work environment will include high-quality human comfort (ergonomics), and improved visual quality.

There will be an increase in the quality of public structures for institutions such as schools, hospitals, libraries, governmental buildings, and local parks. Due to extended periods of education by more people, educational facilities will be of great concern for the contractor. As the level of affluence of the country grows, people will demand and support higher quality educational facilities. This demand for higher-level educational offerings by the population will make existing educational facilities automated. As a result, there will be an enormous quantity of construction projects to replace substandard schools.

A continued increase of individual income will make personal choice more evident in the construction industry. The biggest proportionate increase in construction projects will be towards recreation-oriented facilities, such as water sports, golf, spectator sports, restaurants, hotels, other entertaining facilities, and second homes.

The investment in residences will continue to grow and will provide both increased size and amenity. As income increases, there will be a desire for more space per person. This trend will continue for years to come. The apartment will continue to be a symbol of independence for the young, the single, the single parent, and the urbane. The individual family home will continue to be the ambition of the majority of people.

Contractors. The future contractor will continue to direct the construction project, and the work that cannot be subcontracted. The majority of their efforts will be expended upon bidding, scheduling, and directing the sequence of operations.

The method used to direct construction will vary in the future with the contractor moving more in the direction of off-site fabrication which will reduce the amount of field direction. The trend in off-site fabrication will continue toward pre-assembled component systems with the majority of the direction occurring in a factory. On-site construction will continue to be performed by individual subcontractors. However, the trend will continue toward subcontractors who are system installers that coordinate several trades.

In the future, responsibilities of contractors will be more diffused. It will no longer be possible to find contractors with only a few roles or with a particular type of enterprise. Contractors will be required to practice in an expanded range of roles as described earlier — some as specialists in one or two activities, others as broader generalists with a wider range of capabilities, but none will be considered as a master of all.

Training. In general, the future will require higher levels of expertise in both social relationships and technological processes and less emphasis on the skilled crafter. There are indications that the construction industry, through educational institutions, is beginning to prepare people to better control and create an improved physical environment. Many of those studying construction are combining their study with other fields and will select a role or activity not traditionally associated with the contracting profession. The demand for these persons will be greater than their numbers can provide. The majority of the future graduates will be unlike those of a generation ago and will bring new and more advanced skills and knowledge into the profession. Future contractors will need to develop competencies in the following areas:

1. The understanding of the environment, human needs, and the social criteria to form a theoretical base for developing safe and meaningful environments.

2. The extended design process, which includes the formulation of techniques for syntheses of both social and technological advances through new techniques.

3. The design, development, construction, and implementation of technological and management techniques for greater economic efficiency for delivering structures to a larger percentage of the population.

Courses in these areas will become necessary for advancing the expertise of the contractor and the level of achievement of the construction industry as a whole.

Educational institutions must continue to change, grow, and develop. The improvement of the total environment depends on how successful educational institutions are in recruiting creative persons into the construction program. Attracting the most innovative minds to study, learn, practice, and participate at all levels in the construction industry is crucial. It is equally important to realize the potential of the most able persons outside the field to participate in professional services to develop effective collaborations.

CONCLUSION

Society has used the construction industry to build a civilization. All of us share the idea of living graciously and safely in ways that insure no one is deprived of food, clothing, or shelter. The task of the construction industry is to produce agreeable and efficient structures in pleasant surroundings, that produce an environmentally sound and safe world. Construction projects will be seen as a healthy, normal process.

The constructor of the future must recognize that they are an integral part of the total physical and social environment and that construction will concern the world rather than just the U.S. Constructors must also recognize that we are in a technological age and that society will demand the highest achievable environmental quality. Thus, they must develop more productive relationships with an increased variety of professionals and interest groups to promote a joint effort by the construction industry.

Constructors will continue to face the problem of pressure to build environmentally safe structures. More importantly, they will find themselves confronted with the problem of increased competition from industries that have become more efficient in resource minimization methods while improving the environment.

The purpose of this chapter has been to reflect on the past, examine the present and forecast the future of the construction industry. However, the question still remains: Will the industry respond and make needed changes that have a good potential for success? Most constructors will probably continue in the present direction and will spend more time trying to find reasons why the recommendations will not succeed instead of looking for ways to overcome many of the industry's problems. There will be a few who are more imaginative and will perceive how these recommendations will improve the social and physical environment. Hopefully, their success will be accepted by other constructors.

REFERENCES

Barrier, M. (1992). *Total quality management.* New York: Nations Business.

Brown, L.R. (1993). *State of the world.* A world watch institute report on progress toward a sustainable society. New York: Norton.

Camesasca, E. (1921). *History of the house.* New York: Putnam's Son's.

Christie, G.A. (1992). Out of the tunnel. *Costs & Trends* (Southern edition), Mid-year Edition, (pp. 13–17).

Dodge, F.W. (1992). Costs & Trends Project report. *Costs & Trends* (Southern edition), Mid-year Edition, (pp. 23–48).

Dodge, F.W., & Sweet, J. (1992). *Market Trends.* A Report by the Dodge/Sweet's Company. Dallas, TX: Dodge/Sweet's.

Ewald, J., & Williams, E. (1970). *Creating the human environment.* American Institute of Architecture, Urbana, IL: University of Illinois Press.

Gapay, L.L. (1993). Pen-based computers. *Constructor,* 75(1), 32–35.

Hillebrandt, P.M. (1985). *Economic theory and the construction industry.* New York: Macmillan.

Horn, C.H. (1992). Environmental catastrophes. *Constructor,* 74(3), 49–45.

Kerisel, J. (1987). *Down to earth foundations past and present: The invisible art of the builder.* Boston: A.A. Balkema & Rotterdam.

Lefkoe, M. R. (1980). *The crisis in construction.* Washington, DC: The Bureau of National Affairs, U.S. Government Printing Office.

McNiel, S. (1992). The National Association of Women in Construction. *Cost & Trends,* (Southern edition). Year's End Edition (p. 12).

Ross, D. (1989). Construction project planning. *Journal of Industrial Technology,* 5(4), 11–15.

Street, B. (1992). Clubbing together for quality advice. *Chartered Builder,* 4(9), 20–24.

Thomas, L. (1989). Environmental challenges that lie ahead. *American City and County,* 8(3), 17–21.

U.S. Bureau of Census. (1992). *Current construction portfolio.* Washington, DC: U.S. Government Printing Office.

Whitaker, C. (1936). *The story of architecture.* New York: Halcyon House.

Wiseman, S. (1989). How will we grow. *Essays from the National Association of Professional Homebuilders.* (Southern ed.), Dallas, TX: National Association of Professional Homebuilders.

3

Impacts Of Construction Technology

Peter H. Wright, Ed.D.,
Industrial Technology Education Department
Indiana State University, Terre Haute, IN

Construction activities meet one of the most basic human needs—the need for shelter. Because of their central importance to all societies, construction activities are deeply interrelated with all other elements of human cultures and world views. It is impossible to discuss the impacts of construction on society without exploring the impacts of social and cultural beliefs on the designs and methods of construction. Groups of people and the structures they build are inseparable.

HUMANS AND CONSTRUCTION

Construction, like all human activity, can be analyzed in terms of basic human needs. The need for food, shelter, security, clothing, medical care, and so on provide a basis for explaining how and why particular structures are created. When a new factory inspires the construction of new homes and businesses around it, we should not be surprised—the location of the houses provides the occupants with the means to meet their basic needs by their proximity to a source of income and employment. We could then predict that a supermarket would be another major structure created near the new factory because such structures provide people with money, a way to meet their needs.

A house is often the largest single investment a person makes. Similarly, structures represent a solid portion of the total wealth of the U.S. For these two reasons, one would expect that every student would leave high school with a clear understanding of the construction process and of the elements which separate strong, useful structures from shoddy, ill-designed ones.

However, this is not the case. Most citizens of the U.S. leave high school and later make large investments in a residence with no organized instruction and little knowledge about construction. Is it any wonder that many purchase poorly designed and poorly built dwellings? Why is it that we teach our students so little about an area of technological activity that is critical to both our economy and daily survival?

STRUCTURES AND SOCIETIES

Alexander, et al. (1977), in *A Pattern Language*, provided an integrated philosophy of land-use planning and design based on a series of carefully developed precepts. Among their assumptions was that structures should be designed and located so as to maximize the benefits for all members of a society, not just the most powerful. Their entire building philosophy and later compendium of design patterns for structures were based on a clearly elaborated set of underlying patterns and assumptions.

As an example, they concluded that the maximum number of people in any one political unit should be 2–10 million people. From this, and other assumptions, they proposed the design of cities be planned with limits on the size of commercial concentrations and the development of "city-country fingers." These fingers would consist of urban areas which extend out into areas which are, by law, kept green and undeveloped.

Reading these authors' assumptions and conclusions raises the question of what assumptions and conclusions (spoken and unspoken) underlay our current patterns of land-use planning and design. One way to sense our own patterns is to compare our structures with those of other cultures.

Human dwellings in other countries are influenced by the same factors that influence the form of dwellings in the U.S. — available materials, climate, wealth, living patterns, culture, proximity to food and work, and so on. It is the area of cultural determinants which can best be revealed by comparing dwelling patterns in other countries to those in the U.S.

In many African countries, human dwellings consist of clusters of small huts joined together with fences to form a small central courtyard. Each compound then becomes the dwelling place for an entire extended family (Rapoport, 1969, p.56). The grandparents, parents, and grandchildren all share the same living place, food, and work. In this system, the work usually consists of managing the family's growing plot and/or livestock and of building and maintaining the structures in the family compound. All of these are usually located within easy walking distance.

In these communities, the nature of the dwelling design reflects the societal custom of large extended families living together. The grandparents help with cooking and childcare and there is almost always someone at home in the compound to guard the family's possessions. However, most Americans would find it quite inconvenient and restrictive to live with all of their immediate relatives.

The American ideal of dwelling is the single-family home. In this ideal, the family consists of a nuclear family of parents and their children. It is not uncommon for a builder in the U.S. to turn 60 acres of farmland into a subdivision containing 100 very similar detached single-family homes, each with its own foundation, utilities, landscaping, and access roads.

These subdivisions are usually located a significant distance away from work and school. Unlike the African villages, there is no on-site food-production capability. As a result, the subdivision communities are largely uninhabited on weekdays, during the daylight hours. Residents often spend 1–3 hours per day travelling to and from their homes. The planning of such a community assumes 100% car ownership by all adults who wish to function effectively. The separation of daily activities and the large distances involved require private cars.

The Role Of Public Buildings

In any society, it is possible to analyze what is truly important to the people in power by studying the largest public structures. For example, in medieval Europe, all of the largest most impressive structures are cathedrals and castles. Religious expression and self-defense were clearly the most important priorities.

In ancient Egypt, the pyramids were the largest public structures. They were built as tombs for the pharaohs who were considered divine. Similarly, our massive publicly-built power plants and dams; ubiquitous public schools, malls, and churches; space shuttles; underground nuclear control centers; and football stadiums provide an interested observer insights into our key public values.

ECONOMIC INTERRELATIONSHIPS

The impact of construction spending on the economy of the U.S. is massive. According to the Bureau of Labor Statistics (1992), there were 4.7 million Americans directly employed in construction-related jobs in 1991 out of a total employed workforce of 108.9 million. According to these figures, one of every 23 American jobs was directly involved with construction (U.S. Department of Commerce, 1992, p. 405).

Similarly, out of a total retail trade of 1,821 billion dollars in 1991, lumber, construction materials, plumbing supplies and other hardware was 92.7

billion dollars. In 1991, wholesale sales of durable goods (intended to last over 2 years) was 846.5 billion dollars. Of that amount, 94 billion dollars was sold in the areas of lumber, hardware, and other construction supplies (U.S. Department of Commerce, 1992, p. 772). This only represents materials sold to other businesses such as contractors and professional remodelers. It is obvious, that if people in the U.S. ever stopped constructing structures, there would be massive unemployment and, at the least, a major recession.

The overall health of the American economy strongly influences construction spending. During the 1980s, there was a period of expansion in the U.S. economy which, combined with very favorable tax treatment of real-estate investment, lead to a boom in construction, particularly the construction of commercial real estate. These buildings consisted largely of high-rise office towers. Similarly, malls, warehouses, and low-rise suburban office buildings were built in far greater numbers than the market was able to fill (Barsky, 1992).

In fact, when the tax laws were revised and the slowdown of the late 1980s and early 1990s occurred, many cities had a glut of office space. Cities such as Houston, at one time, had up to 30% of their office space vacant (Barsky, 1992). This led to the collapse of many property development companies and banks. Looking back, it is not hard to see how federal, fiscal, and tax policies caused a boom in commercial construction which led to a reverse impact of an excess of buildings hurting the economy of many U.S. urban areas.

In 1992, one real-estate consultant estimated that at current utilization rates, the U.S. would take 13.5 years to fill (i.e., rent) all of the existing office space available in American cities if no more office buildings were constructed. A problem for building owners is that as tenant leases become due, the tenants have a choice of either moving into another building at a lower price or staying and negotiating a lower price with their existing landlord. Thus, the office space overhang is causing office rents to continue dropping in many cities. This trend is a boon for tenants but it threatens the financial stability of organizations which own and hold mortgages on large office buildings (Barsky, 1992).

Similarly, the U.S. Bureau of Labor Statistics estimated that employment in construction fell from 5.1 million in 1990 to 4.7 million in 1991. Construction jobs are very sensitive to the overall economy and these are not just numbers. In one year between 1990 and 1991, the U.S. government estimated that one of every ten construction jobs vanished (U.S. Department of Commerce, 1992, p. 405). This caused a major strain on workers, contractors, and sellers of construction materials.

Construction has a major beneficial impact on the total wealth of any country because well-constructed structures remain useful longer than the products of most other industries. The average car may last ten years, but the average structure can easily last 50 years. Structures and the land they rest on make up a substantial portion of the total wealth of the U.S. (U.S. Department of Commerce, 1992, p. 464).

In addition, one should not underestimate the economic impact of a newly constructed highway interchange, airport, or dam on the local residents. When the METRO stops were completed on the new Washington, DC subway system, property values near each stop increased dramatically. Furthermore, patterns of construction, retail activity, and business locations were altered permanently. Similarly, the developer Olympia and York was so convinced that a subway link to the London Underground would save its monstrous investment in its troubled Canary Wharf Project, they offered to pay $687 million to build the link themselves (*New York Times,* 1992). Infrastructure construction (e.g., roads, bridges, power lines, and sewers) also allows other industrial activity and wealth creation that would otherwise be impossible.

ENVIRONMENTAL IMPACTS OF CONSTRUCTION

Ironically, most structures are built to create improved environments for humans. What is shelter except for an improved environment for people to sleep, work, and live as compared to living outdoors? Nevertheless, the improved interior environments created through construction have major, and often unfavorable, impacts on the exterior environment around them. The environmental impacts of construction systems can be divided into two main components:

1. the impacts of the structures

2. the impacts of the construction processes themselves.

Impacts Of Structures

Physically, any completed permanent structure drastically affects its immediate physical environment. It alters the drainage, wind patterns, vegetation, wildlife, solar distribution, and human, light, and vehicular traffic patterns.

As an example, these effects can be especially pronounced if a large multi-story structure is built across the street from your house. You may notice that the building blocks the sunlight for hours each day and that the wind accelerates in the cavities around the new skyscraper. The temperature outside your house may vary 10 degrees hotter or cooler under certain conditions than it did under those same conditions before the structure was built. In addition, foot and vehicular traffic will be greatly increased and you may notice changes in other areas of your life such as drainage, noise, shopping opportunities, and water pressure.

Larger projects such as power plants, large factories, and major airports tend to have even greater environmental impacts on the surrounding areas. For these reasons, any group proposing a large construction project is required to complete an environmental impact statement (EIS) which details the expected impacts of the project on the physical, social, aesthetic, and economic environments. Often, financial benefits of the proposed project must be balanced against the inevitable environmental disruptions that result from nearly all construction activities.

All significant impacts in areas such as air and water quality, noise, solid waste, radiation, hazardous substances, vegetation and wildlife, energy supply, natural hazards, and land-use effects (recreational, historical, aesthetic, and socioeconomic) must be covered in the EIS (Rau & Wooten, 1980). Different types of projects involve different areas of emphasis. For example, the impacts of airport construction projects "generally focus on five major areas: noise, air quality, water quality, social impacts, and induced socio-economic impacts" (Rau & Wooten, 1980, pp. 1–29).

The interiors of structures also have multiple impacts on their occupants. One area that has become increasingly noticed is that of indoor air quality. Many houses and office buildings have levels of toxins in the air far higher than that outside. People with multiple chemical sensitivities (MCS) are extremely sensitive to the presence of chemicals such as benzene and formaldehyde in the air. These chemicals, and others, are commonly outgassed by construction materials such as treated fabrics, glues, and new carpets (Beach, 1989). Outgassing refers to the process by which certain products release gases from chemicals used in the manufacturing process, particularly when the products are new.

The Environmental Health Center in Dallas specializes in treating people with MCS. They have treated 17,000 people in the last 12 years. John Bower of Bloomington, IN builds non-toxic homes. Most are bought by people with MCS. People with this condition need the absence of chemical agents provided by the hypoallergenic materials Mr. Bower uses. He avoids carpet and plywood and uses steel studs. These materials add $5–10 to the cost per square foot of new houses (Bower, 1989).

The condition of MCS is becoming more and more recognized. Symptoms of MCS can include tension, memory loss, fatigue, headache, and depression. Wood-products manufacturers are already making progress in a program to reduce the outgassing of potentially dangerous substances from their products. Within 10 years, all housing component manufacturers will have embarked on similar programs.

Impacts Of The Construction Process Itself

If you live or work next to a project under construction, you may notice an increase in heavy traffic and noise. Less obvious may be the changes in vegetation, microclimate, and wildlife. This author lived in a built-up town a few hundred yards from an overgrown field which was suddenly cleared and bulldozed to erect a small commercial complex. Suddenly, the whole neighborhood was overrun with raccoons, rats, rabbits, and other small animals fleeing their former living places. This definitely affected our lives for several weeks. Obviously, the animals were more affected than us!

Many residential construction projects used to result in large amounts of soil runoff from the areas of the project that were excavated or bulldozed in the construction processes. The eroded soil often clogged small streams and other drainage systems. Enacting of new laws have made contractors make a greater effort to control these problems. One economic impact farmers near a new subdivision development may discover is that contractors appear in the fields offering to buy numerous bales of hay to help control soil erosion from their construction sites.

Construction activities indirectly affect people and environments hundreds of miles away through their use of materials. The lumber and plywood industry of certain regions of the U.S. is affected daily by trends in construction activity. Similarly, concrete factories, shingle shippers, building materials suppliers, and the people working in stone quarries are all strongly affected by distant construction activities.

POLITICAL RELATIONSHIPS AND CONSTRUCTION

People in the U.S. have a strong interest in personal freedom and private property. Originally, this resulted in laws protecting peoples' rights to do whatever they wanted with the property they owned. As populations became more dense and more powerful technologies began to be used, voters in many areas became concerned about total anarchy in land use. This lack of regulation resulted in auto-salvage yards operating next to schools and large, dirty, noisy industries being constructed next to quiet, established residential areas.

Over time, many communities developed zoning laws which were designed to plan the locations of different types of land use within their communities. In this way areas were set aside for single-family dwellings, multi-family homes, light industry, heavy industry, and so on. Similarly, building codes were developed so that structures built in the community could be guaranteed to meet certain basic standards of health, safety, and quality. Such ideas were not new. Comprehensive city planning was developed by the ancient Egyptians and later was adopted by the Greeks and Romans (Alexander, 1979).

The interest of the public and private rights are frequently in conflict during the planning and construction of large public structures such as highways, dams, and airports. To build a continuous structure such as a highway, it is necessary to own all of the land in the path of the road. If even one seller refuses to sell or demands an outrageous sum for their land, the project can be stopped entirely.

Laws of "eminent domain" allow governmental bodies to seize the property they need to complete large construction projects if such projects are approved by using proper procedures. Land owners can then accept the amount of money being offered them for their property or fight the proposed monetary offer in court. They cannot, however, choose to keep their land.

While this may seem, and often is, very harsh for the people whose land is being seized, we would have few, if any, new roads or airports without the eminent-domain process. For example, it is hard for most of us to imagine living without the interstate highway system. This collection of highways would never have been built without the eminent-domain process.

Another political ramification of construction-related decisions is that certain technological choices imply certain political and social structures to support them. For example, houses with electricity supplied by individually owned solar panels would imply a system of necessary governmental controls of energy that was unobtrusive. People would create much of their own power and the government would only have to regulate how individuals were connected to the central electrical grid.

This example implies political relations which are far different from the strong hierarchical security system necessary to provide safety from disruption for an electrical system based on large, centralized nuclear reactors. Gorz (1980) pointed out that the choice of nuclear power can be made safe technologically, but only if we assume strong security at the reactor site. A cadre of technical and security elites providing an array of secret training programs, security checks, surveillance systems, and so on is necessary to insure the proper operation of what Gorz called the "nuclear knighthood." Tight, organized security, of course, can be provided but only at the cost of greater secrecy and centralization of power in any society.

With all organized governments, there is a constant struggle between the rights of individuals to do what they want and the rights of the larger community to arrange rules to benefit the greatest number of people, even when those rules restrict individual freedom in some ways. Such a debate is alive and well in the area of construction.

Several west-coast states have passed laws banning all new construction within so many feet of the ocean. Many other states have established statewide commissions which severely restrict such activities. The Federal Wetlands Act, restrictive sewage tap-on policies, and large-lot zoning can also serve to tightly limit what people can build on property they own. On the other hand, in some rural counties of the U.S. there are no zoning restrictions and no building codes.

If the state passes a law which effectively prevents you from building a house on land which you bought for a house, does the state then owe you the same "just compensation" as that owed people whose land is seized under eminent-domain laws? This happened to Mr. Lucas in South Carolina after he paid nearly a million dollars for two beachfront lots which were later rendered unbuildable by South Carolina's new beachfront-protection law. In 1992, a federal appeals court ruled that Mr. Lucas was owed reimbursement from the State (Carlton, 1992). Obviously, this ruling raises far more questions than it answers.

Currently, the trend in the U.S. is toward increasing governmental controls on what individuals can construct on their privately owned property. While this additional regulation often serves to protect people from the actions of others, it must be remembered that there is also a cost. Each additional regulation and approval process increases the total cost of constructing a given structure without adding anything to its final value (assuming it was safe and well planned to begin with).

Construction And Safety

Construction activities are inherently dangerous. Often conducted outdoors in a changing climate, the nature of construction causes the job site to change every day. In addition, numerous workers with different skills and responsibilities work near each other with each group attempting to complete their tasks as quickly and accurately as possible. Accident rates for construction workers are well above the national average for other occupations.

Federal and state Occupational Safety and Health Administrations (OSHA) devote a significant amount of energy to regulate and inspect construction activities. Building codes, carefully controlled union apprenticeship programs for craft workers, and stringent safety regulations all represent attempts to reduce the dangers in the construction process. Nevertheless, on-site construction, due to its nature, will probably always be a more dangerous occupation than most others.

In any event, the safety risk of construction activities is far outweighed by the safety benefits of having decent roads, bridges, houses, and other structures to use. Without construction, we would all be in grave danger every time we lay down in the forest to sleep!

Technological Developments And Construction

Technological innovations have had a great impact on the history of construction systems. The development of the skyscraper in the late 19th century reflects this type of process. Within a span of 20 years after 1850, the Bessemer process permitted significant expansion in the production of high quality steel. During this same period, the Otis elevator, advances in water supplies and sanitation, and the popularization of Portland cement provided the additional elements to make skyscrapers feasible. Due to social and economic trends which lead to such phenomena as overcrowding in urban centers, multi-story buildings were being constructed in every large city.

However, most of the steel being used in construction in the 1870s was not being used in buildings, but rather in railroads. Twenty years of canal construction ended abruptly with the start of a massive era of railroad construction. This railroad boom was made possible by the steam engine along with the Bessemer steel process mentioned above. Railroad construction, in turn, led to social and economic impacts including the full development of western America and a massive immigration of Chinese railroad workers into the U.S.

CONSTRUCTION AND OTHER SYSTEMS OF TECHNOLOGY

There are many connections between construction systems and the other major types of technological systems—manufacturing, transportation, and communication. Developments in each of these areas influence both the design of structures and the methods by which they are constructed. Similarly, construction decisions have impacts on the development of the other system areas.

Manufacturing Systems

Construction and manufacturing systems are closely related because both represent production processes (i.e., those that result in a product). Most construction materials are manufactured rather than just processed or collected. In fact, the amount of manufactured products used in construction processes has been increasing annually.

As an example, doors were often manufactured off site in the past, but each door frame was hand-constructed on-site by carpenters. Currently, a large majority of doors and windows are sold in pre-constructed frames which are just nailed into place on site. A similar pattern of replacing on-site construction with prefabricated units can be seen in the trends towards panelized and sectional houses, prefabricated plumbing trees, modular kitchen systems, and so on.

Transportation Systems

Construction and transportation systems are also totally interrelated. Most transportation systems require massive construction investments in order to function. These include highways, harbors, airports, railroads, pipelines, and electrical transmission systems. The most famous example of these links is the interstate highway system which was started in 1958.

The resulting highways around major American cities permitted many people to commute to work in their cars from previously undeveloped suburban areas. The resulting migration made many real-estate developers wealthy. After many people with money moved out of our cities, a decline in the commitment to universal public transportation and state-of-the-art schools became apparent in many of these same cities. Construction and transportation systems interacted in this example with unforeseeable consequences.

Communication Systems

Construction and communication systems are also interrelated in multiple ways. There is a trend in the U.S. for people to demand greater and greater communication capabilities in their homes, offices, and factories. For example, many people will not buy a house that cannot be connected to cable television. Similarly, it is common for factories which are part of large corporations to have a satellite dish on their roofs to facilitate communication with the home office.

Office buildings have been affected to an even greater extent. Many new buildings have raised floors to permit computer, phone, and power cables to be connected from any point under the floor. Older buildings without built-in data pathways have been rewired, and even torn down, if rewiring proves prohibitively expensive.

Computers, and other information-management technologies, have similarly had a great impact on how structures are constructed. With all of the paperwork, regulation, and complexity of any large construction project, it is safe to say that far fewer structures would be constructed without the legions of computers and the piles of specialized software used by contractors, construction material suppliers, and developers.

FUTURE TRENDS IN CONSTRUCTION SYSTEMS

Certain trends in construction systems and their impacts in the U.S. seem to be occurring and will probably continue. They include:

- the incorporation of more information and less mass and energy in new structures;

- an increasing variety and specialization of structures and uses;

- increasing paperwork, regulation, and litigation;

- more prefabrication of structure's components; and

- more integration of shelter functions.

More Information And Less Mass

In his important book, *The Next Economy*, Hawken (1984) predicted that future products would contain more information and less mass than current versions. Certainly, today's manufactured products have more complex designs, more product changes, and less total materials than earlier versions.

A typical new house made of studs today or a typical small commercial building will weigh less than those constructed with similar systems in 1950. Less materials are used and modern truss and roof systems are relatively light. However, the more important side of the equation is that new structures contain significantly more information than their predecessors in the form of complex materials and designs. They will also be capable of handling more information.

Most new subdivisions in the 1990s will not contain a number of identical houses like those of the 1960s. Due to Computer Aided Drafting (CAD) systems and a greater variety of standardized components, it is easy to build a variety of homes of similar cost but different designs in the same residential cluster. Each house contains the information of its own unique design.

Much has been written about "smart buildings" which permit greater information management within a structure. The application of computer chips to inhabited structures involves the use of sensors and control devices which permit occupants to remotely monitor and control almost every electrically powered system in the home or office. Security systems, heating and cooling systems, and cooking are a few of the areas where major advances have been demonstrated.

Less Energy Use

The elements of passive solar design for both heating and cooling are relatively simple. In northern climates, it is only necessary to orient windows mainly to the south and provide them with an overhang to block the summer sun. Similarly, in desert areas, northern exposures aid in cooling. Passive solar orientation of new structures will soon become the rule wherever possible. New window coatings provide numerous energy and light controlling options for all types of structures (Fisette, 1989).

Similarly, new house construction will continue to become better insulated. Swedish homes are already twice as well insulated as homes in northern Minnesota. The Rocky Mountain Institute estimated that a typical house in Austin, TX could be retrofitted to reduce its energy use by 63% with an economic payback of only three years. This would involve the use of more efficient lighting systems, window treatments, appliances, and insulation (Flavin & Durning, 1988). A study of 40,000 housing energy retrofits in the U.S. showed an average decrease in energy consumption of 25% and a 23% annual return on the cash invested (Flavin & Durning, 1988).

People can and will live well using far less energy. In urban areas, neighborhoods will be more functionally integrated. People will be able to walk to and from work while staying in their neighborhoods. Life-styles such as these decrease energy use and are usually more satisfying than the standard hourly commute. Townhouses will increasingly be built to conserve space and to improve energy efficiency just as they were in the past.

As structures have become tighter, a concern for indoor air quality has increased. No one realized how many potential toxins are released inside a home, particularly a new one. Specialized construction techniques and materials will continue to be utilized to minimize or eliminate outgassing from construction materials.

People afflicted with, or afraid of, MCS will also need space to grow larger gardens to avoid the fear of toxic residues from a variety of chemicals which are routinely sprayed on commercial crops.

Increasingly Specialized Structures

Besides hypoallergenic structures, other specialized structures will be increasingly designed to combine a workplace and a residence. More information will be managed in workplace homes as most of these homes will be wired through computers to central information storage and manipulation utilities. Other homes will be used for cottage industries and increased food production and storage. In fact, the distinction between homes and workplaces may again become blurred as it was in pioneer days.

Certain commercial structures are extremely specialized. The raised floors, computer-controlled central climate systems, sealed windows, and central locations of new office towers make it extremely difficult to convert them to housing or some other use even if they are sitting empty. Older office buildings, while not as useful for modern office equipment, are less specialized and generally easier to convert to housing units.

A specialized house plan which will be employed more often is a plan which provides shared house occupants with two equal, but separate, master-bedroom suites, but shared living areas. A similar plan involves the creation of a ground-floor master suite with other bedrooms located upstairs (Cauley, 1991). This housing theme can be used by the elderly and other unmarried friends. A related design has a small semi-detached apartment attached to a standard single-family home which also provides family-unit options rarely available in existing houses.

As the baby boomers continue to age, many houses will be built or converted to allow people to live without climbing stairs. Similarly, a far greater percentage of new homes will have wheelchair access to all the rooms necessary for life. The information and designs to meet these specialized needs will be readily available thanks, in part, to the automation of the time-consuming chores of the home-design process.

Increasing Paperwork, Regulation, And Litigation

All systems of technology are currently experiencing the trend in the U.S. of increasing paperwork, regulation, and litigation. There are more people, more powerful technological systems, and greater understanding of the complexities of the interrelationships of technological and human systems.

Before people knew that asbestos could help precipitate a form of lung cancer, no one needed a license to remove it from structures and there were no asbestos-liability lawsuits. Now, there is a multi-million dollar history of asbestos-related court awards and stringent EPA-approved procedures for both removing asbestos and for documenting the removal process.

Similarly, in land-use law, the case of Mr. Lucas and his South Carolina beachfront lots discussed above would never have occurred in the 1950s when there were no state laws regulating construction near the ocean. Now, towns and governments who choose to plan for their communities must ask additional questions about how such laws will affect peoples' rights to use their own property.

The Dodds bought 40 acres of Oregon forestland to build a house on in the early 1980s (Carlton, 1992). However, the county soon passed a forest protection law which prevented them from building. Like Mr. Lucas in South Carolina, they sued. However, they were still permitted to cut down the trees. After they sued, a judge ruled that the county did not owe them money because there was not a "total taking" but only a "partial taking" of their property's value (Carlton, 1992). Obviously, decisions such as these can only result in numerous additional regulations, lawsuits, and confusion.

More Prefabrication And Refabrication Of Structures

With prefabrication, materials are imprinted with human design information in an indoor controlled environment. For tasks such as constructing trusses or door frames, work in this controlled environment has proven itself more efficient than similar tasks performed in the field.

Prefabrication of both construction components and entire structures will increase. More and more construction will become assembly-oriented on the job site, and production-oriented in the component factories. Systems such as stress-skin panels, modular-steel commercial construction, and simple sectional homes promise to extend the trend towards the reduction of labor involved in the actual on-site construction process.

Grafton, WV is nowhere near any boom housing construction areas. Still, Ken Auvil, of Grafton Manufactured Homes stated confidently, "We are competing with the stick-builder, but we have better control over our costs." (Cline, 1986, p. C1). His relatively standard, solid, one to three bedroom units sell well in an area with little new construction occurring.

It is important not to ignore the massive store of our national wealth represented by existing residences. Many existing structures are valuable and cannot be recreated. In the future, many of them will be remodeled to incorporate some of the trends described above. Remodeling and rehabilitation of existing structures will grow as a portion of the overall construction market.

More Integration Of All Functions

The new structures of the future will increasingly have their functions carefully integrated. As the trend to functional integration grows, single function rooms will be removed from most shelter designs. An example of the integration of functions which will be common is the shared-use of home information utilities in one area of the home for both entertainment and a home workspace. Careful space management, or even movable partitions, will permit the most efficient use of home audio, video, and data-processing systems.

The combination of telecommunications technology and the draining multi-hour commutes to work around major American cities has led to more telecommuting where people work from their homes and transmit the work digitally over phone lines. Telecommuters then need a room in their homes which can duplicate the functions of an office-work environment. This increases the amount of integration in the home. Visionaries in our culture are also promoting the reintegration of food and energy production in residential and commercial structures. The most obvious example of this trend is the spread of commercial and residential structures utilizing passive-solar heating.

The Todds (1984) proposed that all design should "follow, not oppose, the laws of life" (p. 22). In their work at the New Alchemy Institute, they have produced multi-purpose structures, that can provide enough energy to produce fish and vegetable crops, heat themselves, and shelter people. Similarly, the well-publicized Biosphere II Project is an attempt to integrate buildings with human and food-production systems. Obviously, these types of structures are complicated and time-consuming to manage in their present forms, but they provide a guide to how people may wish to better integrate residential, production, and living systems in the future to enable us to live less destructively in our environments.

Many large, new urban developments are including commercial, office, and residential units in the same development. Developers find that a careful integration of functions strengthens the overall project. For example, having full-time residents around the complex improves the security for office workers who must leave their offices after working late.

CONCLUSION

Construction activities directly meet one of the most basic of human needs—the need for shelter. They also permit us to meet our other basic needs through the production of factories, roads, bridges, and so on. Groups of people and the structures they build are so interconnected as to be inseparable. In fact, one way to understand the culture of ourselves or others is to analyze patterns of public and private construction.

As with other technological activities in the U.S. today, it appears that construction systems continue to use more information and less mass and energy. Similarly, they are subject to increasing regulation and litigation. Future structures will probably respond to social and economic trends by continually integrating a greater variety of functions in single structures. Prefabrication of components will increase.

Construction activities play an important role in every economy and the products of construction make possible most other economic activities. Structures, and the process of their construction, always create impacts on the surrounding environment. Legal and political issues are intimately involved with construction in such areas as community planning, building codes, and the eminent-domain process.

REFERENCES

Alexander, C. (1979). *The timeless way of building.* New York: Oxford University Press.

Alexander, C., Ishikawa, S., & Silverstein, M. (1977). *A pattern language.* New York: Oxford University Press.

Barsky, N. (1992). Manhattan office tower exemplifies one kind of risk: As leases are rolled over, owners have to cope with lower prices. *The Wall Street Journal. 74*(17), B3.

Beach, D. (1989). Learning to deal with chemical sensitivities. *In These Times. 13*(31), 13–14.

Bower, J. (1989). Building healthy houses: Alternative materials can help protect your client from indoor air pollutants. *The Journal of Light Home Construction. 7*(11), 21–23.

Carlton, J. (1992). "Takings" cases don't always favor takers. *The Wall Street Journal. 74*(20), B1.

Cauley, R.M. (1991, January 13). Separate sleeping quarters give families more privacy. *Atlanta Journal,* Section SS, p. 3.

Cline, D. (1986). Outlook bright for Taylor industry. *The Dominion Post.* 12/30/1986, C-1.

Fisette, P. (1989). High performance glazing. *Fine Homebuilding.* (55), 78–81.

Flavin, C., & Durning, A.B. (1988). Building on success: The age of energy efficiency. (Worldwatch Paper 82). Washington, DC: Worldwatch Institute.

Gorz, A. (1980). *Ecology as politics.* (Trans. from French by Cloud, J. & Vigderman, P.). Boston: South End Press.

Hawken, P. (1984). *The next economy.* New York: Ballantine Books.

Langdon, P. (1989). American house design. *Practical Homeowner. 4*(3), 360–67.

New York Times. (1992, October 3). Olympia and New York in offer to get subway link built. Section A, p. 37.

Rapoport, A. (1969). *House form and culture.* [Foundations of Cultural Geography Series.] Englewood Cliffs, NJ: Prentice Hall.

Rau, J.G., & Wooten, D.C. (1980). *Environmental impact analysis handbook.* New York: McGraw-Hill.

Todd, J., & Todd, N.J. (1984). *Bioshelters, ocean arks, and city farming: Ecology as the basis of design.* San Francisco: Sierra Club.

U.S. Bureau of Labor Statistics. (March 1992). *Employment and Earnings Monthly.* Cited in U.S. Department of Commerce. (1992). *Statistical abstract of the United States.* Washington, DC: U.S. Department of Commerce.

U.S. Department of Commerce. (1992). *Statistical abstract of the United States.* Washington, DC: U.S. Department of Commerce.

CHAPTER 4

Construction In Elementary School Technology Education

James J. Kirkwood, Ph.D.,
Department of Industry and Technology
Ball State University, Muncie, IN

THE ELEMENTARY SCHOOL CURRICULUM

Good teachers guide and direct children in gentle and persuasive ways. They don't condemn, they praise; they don't discourage, they encourage; they don't say "impossible," they enable. When caring, sensitive, knowledgeable teachers teach, children learn that all things are possible, and that they will live in a better world—a world which they can help build. Schools must give children every opportunity to prepare themselves for the future—a future that can be only dimly perceived by even the most progressive teacher. One thing is certain: The society into which today's children will live will be increasingly technological. The future belongs to those who are comfortable with technology.

Technology education began in the elementary school. Along with its historical antecedents, technology education figured prominently in the design of elementary and secondary school curriculum innovations. The efforts of those who believed in the value of constructive activities in education—people such as Rousseau, Pestalozzi, Froebel, Herbart, Sheldon and Dewey—affected, forever, the modern elementary school curriculum.

Technology education in the elementary grades differs little from traditional elementary school industrial arts. Elementary school industrial arts programs, unlike their secondary school counterparts, never ventured far from the principles that underlie and inform current technology education curricula. High school industrial arts programs had placed their focus on studying technology, while in the elementary grades industrial arts kept its eye on children and the need they have to understand their world. All of technology education must continue in that tradition.

Development Of Technology Education In The Elementary Grades

Many people look upon construction activities for elementary school children as something new, but it has a long, if somewhat silent, history. As Babcock noted in 1961, "more people are reading and talking about the benefits for children, and as a result, more people are doing something about it" (1961, p. 33). That is increasingly true today.

Babcock captured the excitement and satisfaction that come from engaging children in what he called "construction activities" (he was referring to general industrial arts activities). He was not the first to recognize that children can benefit immensely from manipulative activities in their classroom. The American Council for Industrial Arts Teacher Education succinctly summarized the historical development of technology education in their 30th yearbook (Barella & Wright, Eds., 1981). Borrowing freely from that volume and from Babcock's research in the 10th yearbook (1961), we can trace the development of contemporary thinking about technology education in the elementary grades.

John Amos Comenius, born in the late 16th century, was an influential educator whose "educational innovations [were] spread throughout the continent and, eventually, the world" (Nelson, 1981, p. 21). Two of his educational tenets exist as principles in modern technology education practices. He advocated teaching about the object first, and then about its characteristics. He also advocated learning through experience and investigation. In other words, he believed in providing manipulative, concrete experiences for the learners before proceeding to abstract ideas.

Jean Jacques Rousseau, (1712–1778) a French educational reformer, believed in "natural" education. He relied on the nature of the child for determining relevant areas of study and for finding appropriate methods of teaching. He advocated manual, experiential learning as the best way for children to become all that they are capable of becoming.

Johann Heinrich Pestalozzi had a love for children of the underclass. Taking his cue from Rousseau, Pestalozzi began an experiment with vagrant children of the poor, using music and manipulative teaching techniques. His children developed rapidly and, although Pestalozzi's first school was a financial failure, other educators noted and imitated his methods. Supported with finances from his writing and from his wealthy wife, Pestalozzi opened a home for war orphans in 1798. He used manipulative activities and manual labor to supplement and enhance formal learning.

Friedrich Froebel, known as "the father of kindergarten," studied with Pestalozzi from 1807 to 1809. He used manipulative objects he called "gifts" in teaching kindergarten children. Gifts were three-dimensional geometric shapes as well as natural objects such as sticks, seeds, and stones. He developed "occupations" for children in which they could manipulate and modify the gifts. He gave children a sense of control over their learning environment. Like Pestalozzi and Rousseau before him, Froebel believed that manipulative learning experiences for the young were the means for teaching abstract reasoning. For Froebel, doing preceded knowing.

Johann Herbart, an educational philosopher from Berne, took a somewhat different approach. He was certainly a proponent of handwork. Herbart's emphasis was on using manipulative activities as a method for teaching other disciplines. Although he based his practices in Pestalozzian theory, Herbart differed from Froebel who saw an intrinsic benefit in handwork as a means of training the mind. Today's technology education curriculum designs echo Herbart when they stress the interdisciplinary nature of technology education.

Pestalozzi's theories came to America in 1860. **E.A. Sheldon** was Superintendent of the Oswego Public Schools of New York where he trained his teachers in Pestalozzian methods. He developed a normal school, now the State University of New York at Oswego, and instituted a "practice school" wherein children worked with tools and materials. His ideas spread throughout the country.

John Dewey helped to make school life meaningful for children. In 1897 he published *My Pedagogic Creed* which many educators regard as "the emancipation proclamation of childhood" (Shepard & Ragan, 1992, p. 26). His University Elementary School in Chicago became a laboratory for testing his theory of the "psychology of occupations." Dewey (1899), in *The School and Society*, explained: "by occupation I mean a mode of activity on the part of the child which reproduces or runs parallel to some form of work carried on in social life" (p. 131). He insisted on activities to integrate life experiences with formal learning. "Dewey strongly advocated that truly educative experiences result only when connected to an activity the child is doing independently. Without this connection, some external results may occur, but not education" (Shepard & Ragan, 1992, p. 27). "Dewey meant the use of these occupations as a method of developing the intellect and not primarily of developing manual skills" (Kirkwood, 1968). For example, Dewey would begin a study of woodworking in the elementary school by having the children first plant a tree.

Early in this century, just as now, there were conflicting schools of thought about the relative values of manipulative experiences in the elementary school. While Dewey believed that it served as a method for teaching the standard disciplines, others felt that handwork would, as **Calvin Woodward**, the father of manual training, had said, keep "the boys" in school. They believed in the intrinsic good of handwork as an activity in and of itself. Woodward's disciples did, in fact, come to recognize the value of manual training as not only a discipline but as a method. Stombaugh (1936) stated:

> In the early nineties (1890s) the idea began to develop that manual training should not be an isolated special subject. Instead, consideration should be given to the mutual influences of this subject and the other studies of the school. Bennett, in 1892, told how manual training, when properly taught, could integrate the other studies of the school. (p. 148)

Dean James Russell of Columbia "saw industrial arts as the basis for the elementary school program. He said that manual training, fine arts, domestic art and domestic science should be dropped in favor of the elements of industries, or industrial arts" (Smith, 1981, p. 188).

Frederick G. Bonser of Teachers College, Columbia University, extended the work of Russell. With Russell, Bonser proposed a compromise within the new field of industrial arts. They contended that the discipline served two purposes and proposed that industrial arts was both subject and method. Russell and Bonser's theories were a connection of Herbartian theory with manual training. Their ideas—which organized the elementary school curriculum around the human adaptive processes of food, clothing, and shelter—were implemented widely. Even to this day, their ideas have created much excitement and given much direction to curriculum development in industrial arts. Modern elementary teachers who teach about "community helpers" are an example of Bonser's and Russel's far-reaching influence.

Bonser and his colleague at Teachers College, Columbia University, **Lois Coffey Mossman** clearly prescient of the current curricular emphasis on understanding a common culture, wrote in 1923, "Is there not also a body of experience and knowledge relative to the industrial arts which is of common value to all, regardless of sex or occupation?" (Bonser & Mossman, 1923, p. 20). They argued that industrial arts be accorded the same emphasis as "geography, history, literature and science" (p. 20).

Technology education, in one form or another, has long been an integral part of the modern elementary school curriculum. Trautman (1990) noted that elementary school industrial arts "has been discussed, referenced and used as a base for high school and post-secondary schools for many years" (p. 50). Babcock (1961) sums the value and impact of elementary school industrial arts upon our technology education heritage:

Another and rather surprising and significant fact is that original thoughts and practices in industrial arts began with young children — children of elementary school age. It is surprising that, with this rich background, current practices in elementary industrial arts seem less understood than secondary school industrial arts programs? (p. 7)

Since Babcock's analysis in 1961, schools have continued to improve their curricula. Wirth (1993) argues that as we adopt participative styles of management in post-industrial education, "we need to adopt active, constructivist learning for students as a central goal for schooling . . . . Such efforts are not unlike those that are helping to revitalize American industries and would tap the core values of the American democratic tradition" (p. 366).

Goals And Expectations Of The Elementary School

The ultimate goal of public schools is to prepare our children to live a productive life in the world in which they will live. According to Nannay (1989), "that world is highly technological in nature. Consequently, it becomes imperative that our schools give every student an insight and understanding of the technological aspects of our culture" (p. 3).

Wirth (1993) also believes that if concern for democratic values remains paramount in curriculum planning, we will:

become a learning society at the workplace, in the schools, and in our communities. We could become a society in which the processes of technological change are disciplined by the political wisdom of democracy — a society in which we would adopt only those technologies and social systems that match our best sense of who we are and what we want this society to be. (p. 366)

A major goal is an understanding of culture. We belong to an interactive world that demands a common knowledge base — a collective literacy about our culture. Just as Bonser and Mossman suggested in 1923, cultural literacy in today's technological world must include technological literacy. Paul DeVore has written, "understanding technology is essential to cultural literacy" (Horton, Komacek, Thompson, & Wright, 1991, p. v). Culture, said Ashley Montagu (1958), "is the way of life of a people. It is the people's ideas, sentiments, religion and secular beliefs, its language, tools, pots and pans, its institutions" (p. 31).

Cultural literacy must begin with the schools. "Virtually all citizens in our society share one common experience. At one time or another in their lives, they have attended school" (Beane, Toepfer & Alessi, 1992, p. 13). Peterson (1990) noted that "the beginnings of a successful elementary school Technology Education (sic) program originate with the recognition that technology is an essential element of all cultures and the study of technology is valuable for elementary age children" (1990, p. 48).

"If strangers share very little knowledge," said Hirsch (1987) in *Cultural Literacy*, "their communications must be long and relatively rudimentary" (p. 4). If people share the same basic understanding, communication is short, succinct, and to the point. People must share a current common understanding of society, which, as Hirsch quoted Patterson, "no longer has much to do with white Anglo Saxon Protestants, but with the imperatives of industrial civilization" (1987, p. 10).

Wirth (1993) did not distinguish the elementary schools from the secondary schools as they create a "form of education that could bring high technology and democratic values into creative collaboration" (p. 366). Hirsch also argued for curriculum reform and strongly emphasized that reform must begin in the elementary school:

> Really effective reforms in the teaching of cultural literacy must begin with the earliest grades. Every improvement made in teaching very young children literate background information will have a multiplier effect on later learning, not just by virtue of the information they will gain but also by virtue of the greater motivation for reading and learning they will feel when they actually understand what they have read. (p. 28)

Beach (1990) agrees when he argued that it has always been true that "the bedrock of future cognitive functioning, attitudes, and idiosyncratic viewpoints about the self and the world is laid down during childhood" (p. 41).

The goal of the elementary school is to have children be excited about learning, to be creative in solving problems, and to become aware of their world. The world in which they live is continuing to become more affected by and dependent upon technology. Adults and children no longer have the option to ignore technology: They are affected by it; and they must use it. Peterson (1986) stated, "the aim of the elementary school Technology Education program is to develop a first-hand understanding of the technology that supports daily life" (p. 47).

The Elementary School Curriculum

Is there a "typical" elementary school curriculum? Babcock (1961) noted that "since organizational procedures vary from school to school, and even from classroom to classroom, it is difficult to describe accurately the typical elementary school. It is possible, however, to review trends and common practices" (p. 45). The first place to look is the shelf of textbooks and trade books used in the classroom.

Given the modern propensity for state-adopted texts, the textbook is the single most available and most used curriculum instrument (Shepard & Ragan, 1992, pp. 216–217). Textbooks used in elementary schools contain a wealth of technology education resources. The trade books used as resources in the elementary school classroom and library are another resource for information about technology. All these resources, easily located in any elementary school, include references to manufacturing, communication, transportation, and construction technologies. In particular, Bonser's concept of the study of "shelter" remains as a strong organizing element for early elementary grades. Teachers must plan activities to make abstract technologies meaningful with the help of the textbooks and trade books used in the classroom.

Exactly when, where, and how each child achieves a particular curriculum goal is subject to debate. The movement on national testing waxes and wanes, depending upon whose voice is most prominent. Presently, national standards are receiving attention by the government. "At the National Summit on Mathematics Assessment in April of 1991, people concerned with mathematics education in the United States gathered in Washington, DC, to figure out what it is that every fourth grader should know" (Ohanian, 1992, p. 21). Conversely, the *Curriculum and Evaluation Standards for School Mathematics* (Commission on Standards for School Mathematics, 1989) put control of curriculum design in the hands of the classroom teacher who knows the differences among her children and can best decide when and where and how Daniel and Alice will learn their math facts.

The typical elementary school curriculum is unique to each school and each classroom. A rich variety of books and audio-visual resources supplement the creative energies of the classroom teacher. People are resources too. Nurses, counselors, administrators, resource room teachers, as well as the "special" teachers for fine arts, music, physical education, and technology education can be—when available—excellent instruments for curriculum design and implementation.

Common Experiences In The Elementary School

We expect children to learn the three Rs in the elementary school. These core subjects form the basis for most of children's common experiences and for establishing a common cultural literacy. Given the wide range of experiences, abilities, and personalities that children bring to school, there must be a wide range of activities, resources and teaching methods to accommodate them. And given the need children have for concrete experiences, an activity-oriented approach to learning must be used. For example, an established belief among elementary school teachers of writing is that before children can write they must experience. Although each child's history of practical experience is unique, their experience generally conforms to the common culture of the community and the classroom. Even for a teacher-centered activity, wherein the children participate in an identical activity, the idiosyncratic nature of each child helps to make the experience unique and worth sharing with others. Activity-based writing experiences allow each of the children to confront their own thoughts and feelings and to express their own creative responses.

When children undertake a construction technology education activity they have a common experience that they can and should share with others. In planning their work, they must consult classroom and school resources. For this they have to read, ask questions, and listen. They sometimes must write, telephone, fax, or telecommunicate with people in the world outside the school such as manufacturers of equipment, suppliers of material, chambers of commerce, or data bases. In carrying out a construction technology education experience, children often produce a single product such as a house, a bridge, or a tower. That product is the result of the efforts of a community of workers. Therefore, children must share tools and materials, they must talk and listen to each other, and lend a helping hand to successfully complete a project.

Regarding the common experiences of workers, Hirsch (1987) stated the following:

> The more specialized and technical our civilization becomes, the harder it is for nonspecialists to participate in the decisions that deeply affect their lives. If we do not achieve a literate society, the technicians, with their arcane specialties, will not be able to communicate with us nor we with them. That would contradict the basic principles of democracy and must not be allowed to happen. (p. 31)

Hirsch isn't talking about elementary school pupils developing an esoteric and arcane set of experiences; he is talking about a shared cultural literacy that is technological in nature.

> The idea of literacy has gradually come to include a larger vocabulary of shared scientific and technical knowledge. Especially today, when political decisions in our democracy have an increasingly technical element, our schools should enhance scientific and technical vocabularies. We require not only that ordinary citizens be scientifically literate but that technicians and scientists master the nonscientific literate culture. To explain the implications of their work to others,

experts must be aware of the shared associations in our literate vocabulary and be able to build analogies on those associations. . . . The implications of technological change can become subjects of public discourse—not about technical details, but about the broad issues of the debate. Otherwise we are in danger of falling victims to technological intimidation. (p. 108)

Children and adults are often intimidated by technology. One of the ways technology influences the learning process is the way new technologies affect cognitive strategies, defined as "skills by which learners control their internal processes of attending, learning, and remembering" (Johnston, 1987, p. 7). Kirkwood and Gimblett (1992) emphasized the need to understand technology when they stated:

It is important that our students increase their technological literacy, their ability to cope with technology in their lives. The vitality of our economic and political system depends upon the ability of our students, our future leaders, to understand, form opinions, and make decisions about how to use technology wisely. Technological improvements are changing not only the way we live but also the way we think about our world. New technology demands and creates new thought processes and patterns. To prepare our children for the future we must embrace new technological developments and incorporate them as effective teaching tools. Educational researchers are assessing the effectiveness of electronic learning in all levels of education. (p. 324)

THE ELEMENTARY SCHOOL CHILD

Our most precious resource for the future of mankind is our children. The schools must do all they can to provide the best education possible for each child. Shepard and Ragan (1992) remarked, "schools serve the basic needs and rights of all children. Inversely, schools serve the basic needs and rights of each child" (p. 100).

Characteristics

Educational psychologists often classify the basic needs of children as physical, social-moral, and intellectual.

Physical. Children have a right to be safe from physical harm, to be adequately fed and clothed, and to live in a nurturing environment. Children have a need and a right to physical activity and exercise. Too often, it is the schools that have had to ensure the physical safety and health of their students. A hungry child learns less; a secure and rested child learns more.

A child should not be asked to perform physical tasks that require greater strength, coordination, and endurance than he or she possesses.

Teachers and administrators must consider the physical similarities and differences among the children when they plan for manipulative activities. First graders need smaller furniture and tools than sixth graders. Children progress in acquiring control first of their large muscles then their small muscles. Boys tend to be more aggressive in their play and physical activity but girls lead the way in coordination and body control. According to Shepard and Ragan (1992), "during the years of the elementary schools, girls tend to be significantly ahead of boys in their physical development. Girls tend to develop large and small muscle coordination earlier" (p. 100).

Social-moral. Social-moral development is a phase of growing from immaturity toward maturity.

> Because children must live in a 'sea of human relationships,' they must learn gradually to engage wisely and constructively in the activities of ever-enlarging groups of people. . . . [The school is] the most effective social laboratory because it is possible to structure the school environment so that social development is more systematic. (Shepard & Ragan, 1992, p. 101)

The activity of groups of people is regulated by an evolved system of social and moral values. "Those values begin with respect for self and others and are manifested in the pursuit of learning, a sense of justice, and the capacity for commitment, reliability, and moral courage" (Rowley, 1989, p. 3). The schools have a responsibility to teach, not religious beliefs, but accepted patterns of behavior. These values "begin with respect for self and others and are manifested in the pursuit of learning, a sense of justice and the capacity for commitment, reliability and moral courage" (Rowley, 1989, p. 3). Teachers have an opportunity to model these values as they go about their daily work. Values which are accepted by all people, such as fair play, cooperation, courtesy, and honesty, become important whenever children are involved in cooperative work or play. In the structured learning situation of the classroom, a group construction activity mandates that all children display commonly accepted values.

Intellectual Needs. Benjamin Bloom has noted the crucial nature of intellectual development in the child's early years. He has noted that from conception to age four, children develop 50% of their mature intelligence and that from the ages of four to eight they develop another 20%. If intellectual development proceeds as Bloom contends, then formal schooling, beginning at age five, must be content with influencing less than half of the intellectual development of the child. Bloom argues that variations in the child's environment have little effect after age eight.

To postpone teaching any subject past these early years will make it less meaningful for the maturing adult. To teach something in a manner not suited for the age group is as great an error. Rousseau recognized the need for appropriate activities in writing of the education of Emile. Technology education curriculum innovators have recognized the importance of technology activities for the elementary school child, but have not demonstrated a full appreciation for the profound effect these activities can have on the intellectual development of the child. They have never articulated a planned curriculum that takes advantage of the readiness of the child to learn relevant aspects of technology. The unique nature of each school and the qualifications, aptitude, and understanding of each teacher preclude a fully articulated technology education curriculum for the elementary grades. Curriculum planning for teaching commonly shared technology information, however, must consider the unique personality traits and intellectual and physical abilities of each child.

Hirsch (1987) says that the commonly shared information that the schools need to impart should begin in preschool. "Fifth grade is almost too late. Tenth grade usually *is* too late" (pp. 26–27).

The Potential Of Children

"Teachers are those who are graced with the opportunity to mold precious clay" (Wittich, 1990). The potential of elementary school children is boundless. All too often, the discipline of the classroom destroys the creativity, the energy, and the curiosity of these young individuals. There is comfort in watching children who have the opportunity to move, discuss, make changes, and manipulate materials and their environment. Often, in the open, creative structure of a manipulative technology education activity, teachers can directly view the successes and failures of the children and make use of the "teachable moment." In such a relaxed and open environment, teachers can nourish the curiosity and creativity of their students. Children must be given the opportunity to be problem solvers, to be praised when they are right, to be encouraged when they are wrong, and to be held in respect at all times.

TECHNOLOGY EDUCATION IN THE ELEMENTARY SCHOOL CURRICULUM

No matter how often we are subtly reminded that technology education is usually taught to middle and high school students, we must not ignore elementary school children; they love technology.

For them, technology means solving problems with their minds and building solutions with their hands, not dealing with fancy equipment or hearing about the latest development in industry. To elementary school children, "technology" is an activity where exciting, interesting, and fun things get accomplished, not a big word vaguely associated with lasers and computers, or with the technology of teaching—educational technology.

Technology education is different from educational technology even though the lines between the two overlap. For example, teachers usually use video as an effective method of applying educational technology to the classroom where children are consumers. If, however, children develop their own video programs in a technology education unit, other students might use their product. In the latter case, the children have become producers.

Robert Seidel is chief of the Automated Instructional Systems Technical Area for the US Army Research Institute in Virginia. He argues, in reference to educational technology, that educators do not readily implement innovations in technology into the curriculum. One of the more exciting technologies for the future of education is "virtual reality." Seidel (1993) writes about the use of virtual reality in the classroom:

> In 1968, approximately a generation ago, George B. Leonard prophesied the 'learning dome,' where children could literally enter the world of history or science and interact as they explored these virtual worlds. Virtual environment technology is now on the threshold of making that possible. But if we are to go beyond the demonstration stage, then we will have to turn our collective imaginations to the more intractable, yet potentially more rewarding, stage of pragmatic implementation. (p. 6)

Teachers do not use virtual reality in the elementary classroom because of the expense, yet it is precisely in the elementary school where it could have the most benefit. Teachers have learned to implement their own type of virtual reality by working diligently to connect classroom activities to the real world. Technologies, new and old, help make connections among Leonard's worlds of "history or science." The real world doesn't divide its problems and activities into categories.

Loepp (1992) comments, in the elementary school, "students focus on solutions to problems without knowing which disciplines they are using. As the need arises for introducing mathematics, science, social studies or technology concepts, the teachers facilitate the learning of those concepts as needed" (p. 18). Educators must not hinder elementary school construction technology education activities by adhering to traditional, established disciplines or categories but must allow for and nurture all the creative diversity found among any group of children. More than a decade before conceiving his eminent definition of industrial arts, Bonser said, "the social and liberal elements in the study of the industrial arts are more significant than are the elements involved in the mere manipulation of materials" (1914, p. 28).

A Construction Technology Education Example

The primary grades are not too early to begin construction technology activities. Younger children learn about the housing of their own culture and other cultures by studying communities. In the sample construction project discussed in this section, the technology activity integrated a social studies unit on Africa with a unit on geometry. This technology activity about structure design and construction served as the integrative focus.

The third grade teacher developed the original ideas for the unit based upon her textbooks and upon some trade books. She coordinated the experience because she knew her children and their needs. She and a technology education specialist discussed the goals of the unit. She wanted the children to be exposed to practical examples of plane and solid geometry, to understand how structures were planned and built, and to develop a sensitivity to other cultures, particularly cultures of African people. The two teachers cooperated in the design of a series of construction technology education problems.

Just as Comenius had suggested 300 years ago, activity preceded concept learning. The teachers designed two similar activities about the geometry of buildings. In groups of four, the children undertook the task of building the tallest structure out of waste paper from the copy machine and a small amount of masking tape. They rolled paper tubes, built platforms, and stacked the paper in various ways. Designing and building a tall free-standing structure from soda straws is a problem-solving activity frequently used in technology education (Daiber, Litherland, & Thode, p. 194) and it is no less appropriate in the elementary grades.

The children applied the principles they had learned from the paper-tube structures and applied them to soda-straw structures. When they were finished with both activities, they sat back to analyze each others' work. They discovered that the taller and stronger structures had all implemented, in some way, the shape of a triangle. With some further experimentation and explanation by the classroom teacher they came to some conclusions about triangles and squares. As a follow-up activity the children constructed dowel-rod-frame cubes and tetrahedrons to confirm their conclusions. They found that the tetrahedrons would support nearly half of the classroom encyclopedia set while the cubes collapsed under the weight of one book.

Next, the children read a story about life in an African village where the people live in either round or square houses (Grifalconi, 1986). Realizing the connection with the geometry they had learned, the third graders saw how the village people used triangles and squares to construct the walls and roofs of their homes. This lively story, taken from the folklore of Cameroon, was centered around houses and included many details about the daily life of the people. The narrator is Osa, a child of the village called Taka. Osa begins:

> It was not until I was almost full-grown and left my village that I found our village was like no other. For the men live in square houses, and the women, in round ones! To me this seemed like the natural order of things. . . . "But what is it like?" you ask. I will tell you how it was, and is, for me. (p. 1)

The children heard and read about subjects that disciples of Bonser would recognize as food, shelter, and clothing. They learned, for example, that cassava was a diet staple of the people in the village. Cassava is also known as manioc and tapioca. While the children went about the task of making authentic models of these homes, they cooked tapioca pudding, just as Osa did for her family, making a direct connection with the Cameroon people by using not only their senses of sight and touch but with their senses of taste and smell.

The connection between geometry, the study of a different culture and construction technology education developed as the unit progressed. The interdisciplinary nature of the unit became clear to the teachers and to the children in the culminating activity of constructing the village of Taka. The design phase used experimentation with geometrical shapes made from paper and plastic straws, the construction phase used a direct connection with the people of Africa as they cooked and ate tapioca and built models of the houses.

This activity, like many other activities in the elementary school, was the result of careful planning and the joyful serendipity that occurs when teachers and children are enthusiastically involved in the exciting process of learning.

CONTENT OF CONSTRUCTION TECHNOLOGY

As A Study In And Of Itself

Methods may change but the content of construction technology does not differ from grade level to grade level. The content of construction technology stands as a resource from which to derive educational experiences. Materials, structures, procedures, tools, equipment, and organization are a data base for curriculum development. Teachers can pick and choose from those which suit the needs and capabilities of the classroom experience. Children are quite capable of learning about principles of design, of categorizing the uses of structures, of management techniques, and of assessing environmental consequences of various construction practices. They can also learn about foundations and superstructures, and transportation structures versus communication structures. Because of the need for elementary school children to learn through concrete, manipulative experiences, abstract ideas such as these are best learned in the context of an activity.

Resources for curriculum planning in construction technology abound in technology education textbooks. They describe construction technology in similar terms and in various organizational patterns. They list materials; they discuss tools and machines; they describe techniques; they organize construction technology into categories of large and small building structures, bridge design, road design, communication construction, transportation construction, and so on. The elementary school teacher can consult any of these resources which have been written for middle school and high school students and adapt the material for the elementary school classroom. Text and trade books discuss current affairs. Newsletters, weekly news magazines, and daily newspapers are replete with information about construction technology.

The content may be the same as secondary school construction technology education but in the elementary school it is more conceptual in planning, more manipulative in carrying it out, and more focused on the child and the interdisciplinary nature of the curriculum. It is the rare elementary school teacher who will teach any technology as a subject in and of itself.

Interdisciplinary

The elementary school curriculum is wholeheartedly interdisciplinary. "At the elementary school level, technology education needs to be integrated with other subject areas to be most effective. Technology infuses readily into the regular classroom curriculum, whether it be language arts, science, math, reading or geography" (Daiber, Litherland, & Thode, p. 189).

Metrication of the construction industry is about to be implemented. The Federal Highway Administration, for example, has set "September 30, 1996 as the target date for full conversion of the agency's annual $16 million construction program" (Construction Metrication Council of the National Institute of Building Sciences, 1993, p. 1). Teaching metrics through construction is a hands-on way to experience and internalize the metric system.

The sample unit of round and square houses, discussed above, demonstrated the interdisciplinary nature of technology education activities by integrating language arts, math, and social studies with the study of African housing. The children learned, in a concrete way they will remember, the nature of triangles and squares and of cubes and tetrahedrons. They learned how geometry was applied to structures. They learned how adults and children lived in a village in Cameroon.

The benefits of construction activities are often more subtle. In the round and square house unit, the children listened to the teacher read the story to them. The library then became a resource as the children themselves located the book, *The Village of Round and Square Houses* (Grifalconi, 1986). They studied the words and pored over the award-winning illustrations, each child internalized the experience in his or her own unique way. The children learned how to cooperate and to be honest with each other. As each group evaluated itself on the criteria they and the teachers had established at the beginning of the paper-tube and soda-straw structure activity, they scored themselves on each criterion. The children, however, had learned the values of cooperation and compassion; they never totalled their scores or compared themselves with other groups.

As long as the classroom teacher remains involved in the planning and execution of the activity, the unit will remain interdisciplinary. The intrinsic interest of manipulative experiences involves the children as no abstract study can and automatically spills over into daily reading, writing, and math experiences.

METHODS OF DOING CONSTRUCTION TECHNOLOGY EDUCATION IN THE ELEMENTARY SCHOOL CLASSROOM

Just as there is no typical elementary school curriculum, there is no typical method of doing construction technology in the elementary grades. There have been and will continue to be many successful implementations of technology education whenever imaginative and caring teachers design learning experiences for their students. This section will outline several construction technology activities. In all of the following activities, the goals of construction technology are integrated with other subjects and the children are engaged in interdisciplinary learning.

Teaching Patterns

Most frequently, the elementary school teacher, alone, designs and implements the activity with little or no outside help. In this case, resources are sometimes difficult to locate and obtain. Resourceful teachers often find that parents, especially those who have technical skills, are willing to assist in providing materials and help with the activity in the classroom. Less frequently, a school has the assistance of a technology education teacher who is assigned or who volunteers to implement a construction activity. In that case, the planning and implementation is facilitated by two specialists — one who knows the children intimately and one who knows the technology.

Developing A Construction Technology Education Unit

Unit planning begins when the classroom teacher designs the curriculum for the year. As with all curriculum planning, the teacher is guided by learning that must be accomplished, and by learning that is optional. Out of the plethora of learning experiences, the teacher selects those that would be learned by the children more easily if they were made manipulative, and then sketches out ideas for activities. Sometimes activities are suggested in textbooks and trade books, or by producers and distributors of commercial or non-profit organizational materials. Other sources of ideas are professional journals and magazines and workshops the teacher attended.

Once the idea for an activity is established, the classroom teacher and a resource person collaborate to bring the idea into the classroom as a hands-on learning experience. If that person is a technology education specialist, he or she will be able to identify an activity that not only enables the learning experience to accomplish the classroom teacher's goals, but also teaches about technology. Curriculum design, under this ideal condition of classroom teacher collaborating with a technology education specialist, becomes a process of merging the content of social studies, math, reading, science, or other traditional discipline with the discipline and techniques of technology education.

The two specialists agree on a rough outline for the unit and both prepare materials. The classroom teacher assesses the needs and abilities of the students. A space for the activity is prepared. Books and other resources are scouted and obtained and displayed for the students to use. The technology education specialist prepares materials and collects the necessary tools and equipment. Both teachers are alert for current events that have an impact on the unit. What often appears to be serendipity is the result of the sensitivity the teachers develop to the subject. Unexpected sources and materials "appear." For example, a new song about building bridges might be discovered, or one of the children's relatives might have an unexpected expertise in the history of a local bridge or might be a construction worker or manager. Although it seems unplanned, one or more happy accidents always seem to occur to enrich the unit.

The classroom is set up for the activity and the two teachers work together with the children. The technology specialist provides most of the technical instruction and the classroom teacher sets the direction and tone for the activity. Both teachers are available to assess the impact of the unit on each child, to direct the students in being successful with their work, and to praise or console, and encourage and cajole, as each planned-for or unexpected teachable moment occurs.

SAMPLE UNITS OF CONSTRUCTION IN TECHNOLOGY EDUCATION

Following are a few actual units which serve to illustrate typical goals and activities for construction technology units in the elementary grades.

Community Planning In First Grade (Kirkwood, 1992a)

This activity begins with the children viewing aerial photographs of cities. The teacher asks questions to help the children identify types of buildings such as single- and multi-family residences, factories and governmental buildings, transportation networks, and other salient characteristics. The children are provided an abundance of scrap wood, including a base upon which to glue the smaller pieces. They are asked to describe the shapes of the wood scraps as they work. They are taught to use a handsaw and asked to cut at least one piece for their city. They explain their finished products to each other. That completes the construction phase, but the unit builds on the geometry, measuring, cutting, and hammering skills the children are developing. They next measure, cut lengths of wood, and practice driving nails. They are encouraged to find designs in their nail sculptures and to outline them with copper wire. Some children will drive nails all around their board, creating "insects." Last they build a geoboard by cutting stock to length and driving nails into a pattern glued to the square board they have created. The teacher uses the board to further illustrate geometric shapes and to teach about perimeter and area.

Technology Education Goals. The children will:

• Visualize a city.

• Use tools safely and competently with direct supervision and assistance (e.g., hand saw, cross-cut saw, back saw, coping saw, hammer, rule, and try square).

- Follow drawings and plans.

General Education Goals. The children will:

- Identify governmental, economic, and social functions of a community.

- Work cooperatively.

- Measure with a rule.

- Name solid shapes.

- Increase their knowledge of plane geometry.

- Increase their technical vocabulary.

Activities. The children will be involved in a variety of tasks. Some of these tasks take only a few minutes, others might take two or three periods of concentrated effort. The entire unit could take from two weeks to two months.

- View aerial photographs of cities.

- Children are given a demonstration of hand saws and glue.

- Each child, using scrap wood, designs and builds their own "city."

- Each child measures a square line and cuts a piece of wood to length as a base for a nailing activity.

- Children drive nails into the face of the wood. Some children will drive nails all around the wood for even more interesting results.

- Children weave soft copper wire around nails to enhance shapes.

- Children cut 3/4" x 10" x 10" plywood from 3/4" x 10" stock.

- Following a template, children nail 2" finishing nails in a 1" square pattern into plywood to create a geoboard.

- Children follow teacher or self-generated shapes on the geoboard with rubber bands.

Housing In The Second Grade

A cooperative and long-term effort is to construct a 1/2 scale model house in the classroom. The teacher is aware of the social action group, "Habitat for Humanity" and wants to impart a similar knowledge of both social concern and construction technology to his/her children. A 1/2 scale, take-down model house is constructed by the children in the classroom, using materials prepared by a technology education specialist.

Technology Education Goals. The children will:

- Build one or more systems of a house.

- Follow drawings and plans.

- Identify the systems of a house (e.g., framing, wiring, and plumbing).

- Use tools safely and competently with supervision (e.g., hand saws, hammer, rule, try square, and screwdriver).

General Education Goals. The children will:

- Describe Habitat for Humanity.

- Describe the need for adequate housing.

- Increase their technical vocabulary.

- List the costs of housing.

- List the functions of a house.

- Measure with rulers and measuring tapes.

- Use technical terms to communicate.

- Work cooperatively.

Activities. This unit takes a large amount of time in planning and execution. Similar to the building of a full-scale structure, the teachers must plan for the proper sequence of procedures (e.g., framing precedes sheathing). The complexity of the planning and the techniques of construction necessitate a certain amount of skill in the preparation of materials. Depending upon resources, skill, and outside help, the following activities will be undertaken by the children:

- Children are given demonstrations of tools and materials.

- In groups, children construct a 1/2 scale, one-room house with an open back which can be dismantled and stored for use as a puppet theater, store, reading room, etc.

Simple Machines And Road Construction Equipment In The First Grade (Kirkwood, 1992b)

Children are given the opportunity and instruction to construct small-scale toy bulldozers, road graders, and cranes. These toys illustrate the practical use of simple machines. The dozer blade works as a lever. The grader demonstrates the wheel and axle, the wedge, and the inclined plane.

The crane shows the pulley. Children are given some of the parts; they follow plans to construct and assemble the rest. One toy is constructed by each child. Assignment to groups is based upon the classroom teacher-assessed skill level of each child. Children play constructively with the toys at the culmination of the unit.

Technology Education Goals. The children will:

- Use the following tools safely and competently with direct supervision and assistance (e.g., hand saws, hand drill, hammer, rule, try square, and file).

- Follow drawings and plans.

- Identify the function of a road grader, bulldozer, and crane in road construction.

General Education Goals. The children will be able to:

- Identify and describe simple machines.

- Measure with a rule and try square.

- Name solid shapes.

- Increase their knowledge of plane geometry.

- Increase their technical vocabulary.

Activities. Children are given detailed plans for building either a bulldozer, a road grader, or a crane. They are supplied with many of the parts and given specific instructions and help in cutting, drilling, and shaping other parts.

Communities Throughout History: Upper Elementary Grades

Mankind has gathered together in social groups since earliest recorded time. Communities of people share some basic characteristics throughout the ages. A historical perspective on the development of civilization is evident upon completion of this activity. In groups, the children research, plan, and build a city. They are asked to include as many relevant features as possible. As they work, they compare and contrast the results of their efforts with those of the other groups. Attention is paid to the discipline of community planning.

Technology Education Goals. The children will:

- Consider geography, climate, and materials available to their city.

- Design and build a model of a city to include housing, public versus private facilities, government buildings, health and safety utilities, transportation, construction, communication, manufacturing, energy sources, entertainment, education, food supply, water supply, air supply, and waste disposal.

- Effectively supervise people, time, or work.

- Follow drawings and plans.

- Learn to safely and competently use or increase their skill with supervision (e.g., hand saws, hammer, rule, try square, screw drivers, electric or manual hand drills, and pliers).

- Trace the development of technology through the evolution of cities through history.

General Education Goals. The children will:

- Trace the history of mankind in the increasing sophistication of cities.

- Identify governmental, economic, and social functions of a community.

- Work cooperatively.

Activities. Children will be grouped and given an assignment to describe and provide a visual model of the appearance of a city based on the following:

- Gathering and/or farming culture (historical or aboriginal).

- Mediterranean, about the time of the birth of Christ.

- Renaissance to the industrial revolution or non-western civilization not covered in the other groups.

- Present.

- Future. Consider assigning environments such as earth as it is now (hospitable conditions), a polluted earth (inhospitable conditions), a colony on the moon, or an underwater city.

Bridge Design In Upper Elementary Grades

Using a different approach, upper elementary students use educational technology to study more abstract concepts. In this unit children use the computer program called Bridge Builder (Rutherford, 1990) before constructing a simple truss bridge.

The computer program is described by the publisher as a "construction set for the computer." It is used to develop pre-engineering skills in middle school and high school students. The basic concepts of how engineers think about forces, however, is suitable for elementary school children. For example, the publisher states "engineers represent forces with arrows. The arrow must show both magnitude (M) and direction. Weight (W) is a force directed toward the center of the earth" (Rutherford, 1990, p. 2–1). These sentences accompany simple drawings of a person pushing on a desk and a fish hanging from a scale. Truss bridges are drawn and explained in the text which accompanies the program. Simple drawings illustrate concepts of compression and tension, internal and external forces, failure modes, and geometric stability. The activity can be made simple enough so that upper elementary school students are successful with the activity. The object of the computer game is to design and build the lightest weight, stable, steel bridge that can carry an 80,000 pound truck between two abutments 120 feet apart.

Technology Education Goals. The children will:

- Identify and describe the structural members of a truss.
- Use the computer to solve problems.
- Understand the principles of bridge design.

General Education Goals. The children will:

- Increase their knowledge of plane geometry.
- Increase their technical vocabulary.

Activities. Each child, first, builds a simple model bridge to a specified length from 1/4" square pine stock. A destructive test is applied. Then they are taught how to use the computer program and given independent time to design a successful bridge. A second simple model bridge is then built from the same stock to the same length and a destructive test is again applied.

SUMMARY

Many people look upon construction activities for elementary school children as something new. Technology education began, however, in the elementary school. Technology education in the elementary grades differs little from traditional elementary school industrial arts. Technology education in the elementary grades can trace its development from John Amos Comenius in the early 17th century, who believed in providing manipulative, concrete experiences for learners before proceeding to abstract ideas; to Bonser and Mossman, who argued in the 1920s that industrial arts be accorded the same emphasis as geography, history, and literature. Currently, leaders in the field such as Paul DeVore, express the notion that technology education in the elementary grades is essential to cultural literacy.

Educational psychologists often classify the basic needs of children as physical, social-moral, and intellectual. The ultimate goal of public schools is to prepare our children to live a productive life in the world in which they will live. That world is highly technological, making it imperative that our schools give every student an insight and understanding of the technological aspects of our culture.

It is difficult to accurately describe the typical elementary school since they differ from district to district. It may be said, however that all schools serve the basic needs and rights of all children.

Elementary school children typically love technology when it is an activity where exciting, interesting, and fun things get accomplished. Technology activities serve to integrate other subject matter. When teaching mathematics, science, social studies, or technology concepts, the teachers can facilitate the learning of those concepts with technology education. As Bonser said, "the social and liberal elements in the study of the industrial arts are more significant than are the elements involved in the mere manipulation of materials" (1914, p. 28). Educators must not hinder elementary school construction technology education activities by adhering to traditional, established disciplines or categories but must allow for and nurture all the creative diversity found among any group of children.

The process of planning construction activity units vary from school to school and from teacher to teacher. Each classroom situation depends upon its unique curriculum, and upon the teacher's knowledge of technology, and the available resources. Technology education serves an integral role in the elementary classroom. Limitations imposed by available materials, skills, and knowledge should not prevent the most excellent of learning experiences—learning by doing.

REFERENCES

Babcock, R. (1961). Elementary school industrial arts. In R.P. Norman & R.C. Bohn (Eds.), *Graduate study in industrial arts* (pp. 33–49). Bloomington, IL: McKnight & McKnight.

Barella, R.V., & Wright, R.T. (Eds.). (1981). *An interpretive history of industrial arts.* Bloomington, IL: McKnight & McKnight.

Beach, L. (1990). *Image theory: Decision making in personal and organizational contexts.* New York: John Wiley and Sons.

Beane, J., Toepfer, C. Jr., & Alessi, S. Jr. (1992). *Curriculum planning and development.* Boston: Allyn and Bacon.

Bonser, F.G. (1914). *Fundamental values in industrial education.* New York: Teachers College, Columbia University.

Bonser, F.G., & Mossman, L.C. (1923). *Industrial arts for elementary schools.* New York: Macmillan.

Commission on Standards for School Mathematics. (1989). *Curriculum and evaluation standards for school mathematics.* Reston, VA: National Council of Teachers of Mathematics.

Construction Metrication Council of the National Institute of Building Sciences. (1993). *Metric in Construction.* Washington, DC: U.S. Government Printing Office.

Daiber, R., Litherland, L., & Thode, T. (1991). Implementation of school-based technology education programs. In M. Dyrenfurth & M. Kozak (Eds.), *Technological literacy* (pp. 187–211). Peoria, IL: Glencoe.

Dewey, J. (1899). *The school and society.* University of Chicago Press.

Grifalconi, A. (1986). *The village of round and square houses.* Boston: Little, Brown.

Hirsch, E., Jr. (1987). *Cultural literacy.* Boston: Houghton-Mifflin.

Horton, A., Komacek, S.A., Thompson, B.W., & Wright, P.H. (1991). *Exploring construction systems.* Worcester, MS: Davis.

Johnston, J. (1987). *Electronic learning, from audiotape to videodisc.* Hillsdale, NJ: L. Erlbaum Associates.

Kirkwood, J.J. (1992a). Elementary school math and technology education. *The Technology Teacher, 51*(4), 29–31.

———. (1992b). Teaching children about simple machines. *The Technology Teacher, 52*(1), 11–13.

———. (1968). *Selected readings: Industrial arts for the elementary grades.* Dubuque, IA: Wm. C. Brown.

Kirkwood, J.J., & Gimblett R.H. (1992). Expert systems and weather forecasting in the 4th and 5th grades. *Journal of Computing in Childhood Education, 3*(3/4), 323–333.

Loepp, F. (1992). The relationship of technology education to mathematics, science, and social studies. *Camelback symposium. Critical issues in technology education* (pp. 15–20). Reston, VA: International Technology Education Association.

Metric in construction. (1993). Construction Metrication Council of the National Institute of Building Sciences. Washington, DC: U.S. Government Printing Office.

Montagu, A. (1958). *Education and human relations.* New York: Grove Press.

Nannay, R.W. (1989). A rationale for studying technology education in the K-6 curriculum. *Technology Education for the Elementary School.* TECC Monograph 15.

Nelson, L.P. (1981). Background: The European influence. In R.V. Barella & R.T. Wright (Eds.), *An interpretive history of industrial arts* (pp. 19–47). Bloomington, IL: McKnight & McKnight.

Ohanian, S. (1992). *Garbage pizza, patchwork quilts, and math magic.* New York: W.H. Freeman.

Peterson, R.E. (1986). Elementary school technology education programs. In R.E. Jones & J.R. Wright (Eds.), *Implementing technology education* (pp. 47–69). Encino, CA: Glencoe.

Rowley, J. (1989, Fall). A matter of values. In D. Burleson (Ed.), *News, notes and quotes* (p. 1). Bloomington, IN: Phi Delta Kappa.

Rutherford, B. (1990). *Bridge Builder* [Computer program]. Aberdeen, WA: Shopware Educational Systems.

Seidel, R.J. (1993). Guest editorial. *THE Journal, 20*(6), 6.

Shepard, G. D., & Ragan W. B. (1992). *Modern elementary curriculum.* Ft. Worth, TX: Harcourt Brace Jovanovich.

Smith, D.F. (1981). Industrial arts founded. In R.V. Barella & R.T. Wright (Eds.), *An interpretive history of industrial arts* (pp. 165–204). Bloomington, IL: McKnight & McKnight.

Stombaugh, R. (1936). *A survey of the movements culminating in industrial arts education in secondary schools.* New York: Teachers College, Columbia University.

Trautman, D.K. (1990). Conceptual models for communication in technology education programs at the elementary, middle school and junior high school levels. In J.A. Liedtke (Ed.), *Communication in technology education* (pp. 50–61). Mission Hills, CA: Glencoe/McGraw-Hill.

Wirth, A.G. (1993). Education and work: The choices we face. *Phi Delta Kappa, 74*(5), 361–366.

Wittich, B. (1990). *Technology education: The new basic* [Video tape]. Reston, VA: International Technology Education Association.

CHAPTER 5

Construction In Middle School Technology Education

Richard A. Boser, Ph.D., Industrial Technology Department
and Dennis Gallo, M.S., Center for Mathematics,
Science, and Technology
Illinois State University, Normal, IL

Turning Points (1989), a report on middle schools by the Carnegie Council on Adolescent Development, declared that, "For many youth 10–15 years old, early adolescence offers opportunities to choose a path toward a productive and fulfilling life. For many others, it represents their last best chance to avoid a diminished future" (p. 8). To help young adolescents choose a path towards productive and fulfilling life, Lounsbury and Clark (1990) recommended that middle school programs must be developmentally appropriate and must actively engage youngsters in the learning process. Drawing on an old adage by Dewey, Lounsbury and Clark noted that "The best preparation for tomorrow is effective living today" (p. 134).

Construction programs in technology education can assist students with the development of effective living skills for today and tomorrow by providing opportunities to explore the built environment and the processes and careers that are a part of it. The construction enterprise is an integral part of most economic activity and is one of the world's largest industries. In beginning their career explorations and understanding of the built environment, middle school students need to be aware of the many career opportunities within the construction industry and the types of systems and technologies employed in those enterprises.

However, these are challenging times for dealing with construction in the middle school environment. Traditional ways of organizing and delivering construction-related curricula are yielding to thematic or interdisciplinary approaches to include the integration of mathematics, science, and technology. The purpose of this chapter is to serve as a resource for the development of construction technology education programs that meet the needs of young adolescents in middle schools. This will be accomplished through an examination of the characteristics of learners in their middle school years and the unique role of schooling at these grade levels. From this student-centered perspective, approaches to the exploration of construction in the middle school will be examined.

DEVELOPMENTAL CHARACTERISTICS OF MIDDLE SCHOOL LEARNERS

For many young people, the transition from pre-adolescence to early adolescence is characterized by unprecedented biological and psychological change (Carnegie Council on Adolescent Development, 1989; Kindred, Wolotkiewicz, Mickelson, & Coplein, 1981; Orstein, 1992). For most students, the middle school years are also marked by the beginning of puberty, described by the Carnegie Council on Adolescent Development as "one of the most far-reaching biological upheavals in the life span" (p. 12). Equally dramatic is the young adolescent's growth in cognitive ability. For many, there is a new capacity to think in more abstract and complex ways.

While the learner's chronological age may range from 10 to 14 years, their actual stages of developmental growth may range from late childhood to early adulthood (Kindred, Wolotkiewicz, Mickelson, & Coplein, 1981). Toepfer (1991) characterized this wide-ranging diversity in physical, social, emotional, and intellectual development as the hallmark of the middle school learner. The summary of developmental changes of adolescents outlined below draws from the work of Gilstrap, Bierman, and McKnight (1992), Glatthorn and Spencer (1986), and Rossi and Stokes (1991).

Physical Development

Adolescents experience a marked increase in height, weight, strength, and muscle mass. Sexual maturation is also occurring. These physical changes are characterized by rapid, but uneven growth, glandular and metabolic imbalances, and many nutritional abnormalities. The rapid growth of adolescents affects their level of physical activity; while at one moment they may appear to have boundless energy, they may tire quickly.

Cognitive Development

Most young adolescents are considered to be at the stage of semi-formal operations, as defined by Piaget (1950). At this stage, students are developing the capacity to deal with more abstract and complex ideas. For example, there may be a marked change in student's political thinking as they develop the capacity to reason political issues from the perspective of concepts and principles. Some students, however, have not yet made this transition and are still at the stage of concrete operations, usually associated with late childhood. Still other learners may have advanced to an adult level of cognitive functioning, or formal operations. Therefore, diversity in cognitive development is the rule among middle school learners. Characteristics of this age group may include: (1) varied learning styles, (2) distractibility and short attention span, (3) broad interests that may often be short-lived and unfocused, and (4) a present-time orientation.

Psycho-Social Development

Adolescent psycho-social development is characterized by the search for independence and acceptance. Young adolescents typically start to loosen the childhood ties and, while still feeling positive about their parents, they begin to act more autonomously and assertively within the family. Moreover, crowds and cliques become important as adolescents seek the acceptance of their peers. Identity formation becomes a social issue as the peer group is used to explore the self, test values and new identities, and get feedback on their effects. During this time of identity formation, the rising middle grade student is also developing an orientation toward achievement and commitment.

Schooling And The Needs Of The Students

Middle schools deal with youngsters at a critical stage in their development. The Illinois State Board of Education (1992a) noted that:

> Adolescents have a need for intimacy, autonomy, and recognition of their individuality. Instead, they are frequently subjected to an impersonal school environment, little opportunity for making their own decisions, and rigid structure which does not encourage variability and the development of individual strengths. (p. 2)

On a national level, the Carnegie Council on Adolescent Development (1989) declared that "a volatile mismatch exists between the organization and curriculum of middle schools, and the intellectual, emotional, and interpersonal needs of young adolescents" (p. 32). Both of these reports suggested that if the needs of middle school students go unmet, these youngsters are more likely to engage in risk-taking behavior, such as experimentation with alcohol, smoking, drugs, or sex, which increase the likelihood of personal difficulties and school failure (Illinois State Board of Education, 1992a).

Therefore, early adolescents need teachers who are sensitive to their developmental changes and programs that are flexible and responsive to their needs. To counter the rigid departmentalized nature of many junior high and middle schools, leaders are moving toward more adaptable school structures. Epstein (1990) suggested that schools dedicated to early adolescents will increasingly adopt (1) flexible schedules, (2) more exploratory courses and mini-courses, (3) the use of interdisciplinary teams with common planning periods for teachers, (4) meaningful parental involvement, (5) advisory programs, (6) cooperative learning, and (7) ways of developing long-term teacher-student relationships.

Technology education has provided middle school students with courses that are exploratory in nature, that provide a cooperative learning environment, and that demonstrate interdisciplinary connections between a variety of academic areas. Unfortunately, these strategies have often been applied internally, only within technology education courses. With the growing trend toward interdisciplinary instruction, technology educators need to find ways to connect technology activities across subject areas.

CHARACTERISTICS OF EFFECTIVE MIDDLE SCHOOLS

What does an effective middle school look like? The Carnegie Council on Adolescent Development (1989) recommended the creation of small stable learning communities, such as schools-within-schools, with a core academic program that integrates the subject matter disciplines; and promotes active learning, critical thinking, and healthy living in order to prepare students to fulfill their obligations as responsible citizens. The Carnegie Council further recommended that effective middle schools are characterized by cooperative learning, the meaningful involvement of families in the education of their children, extensive connections between the school and the community, and the elimination of tracking.

This far-reaching prescription for reform of the middle schools recommended by the Carnegie Council on Adolescent Development (1989) seems to be accepted in principle but as of yet does not appear to be widely realized. A study by the Illinois State Board of Education (1992a) indicated that "of the ten concepts designated as key to middle-level school success, eight are fully implemented by less than one quarter of the responding schools" (p. 2).

However, some progress toward reform is being made. One of the key recommendations for reform of middle school instruction is the need for students to be actively involved in interdisciplinary or thematic activities

(Carnegie Council on Adolescent Development, 1989; Illinois State Board of Education, 1992b; Lounsbury & Clark, 1990; Toepfer, 1991). The Carnegie Council on Adolescent Development (1989) noted that "young adolescents demonstrate an ability to grapple with complexity, think critically, and deal with information as parts of systems rather than as isolated, disconnected facts" (p. 47). Toepfer (1991) suggested that although students are developing these higher level cognitive abilities, the majority of students at this level cannot readily integrate the information learned across separate subject areas. Therefore, students need assistance to make those connections. Toepfer argued that thematic instruction and using interdisciplinary working groups helps middle school students make these connections.

Moreover, for thematic instruction to be effective, it must deal with content that is relevant to young adolescents. Lounsbury and Clark (1990) stated that:

> Educators must make a concerted effort to balance it [traditional discipline based content] with more contemporary content that deals with the real issues that face all youth at this time of life in today's society. The school's curriculum is seen by students as a thing apart—tolerated but not of immediate assistance in their lives. Only when students sense inherent value in the content and activities will they commit themselves to quality efforts. (pp. 136–137)

Although the recommendation for the development of effective middle schools are far-reaching, technology education programs appear to be well aligned with many of the proposed changes. Instructional activities tend to emphasize the broad exploration of problems through active learning. The systems approach advocated in technology education curriculum models also helps students get a handle on the "big picture" and provides students with ways of linking ideas and information. Moreover, the focus on problem solving and technological systems encourages the development of interdisciplinary connections. Through active participation in technology learning activities, students may begin to develop the knowledge to help them understand their technological society and the self-knowledge that will assist them in choosing the path to a productive and rewarding future.

Roles And Functions Of Middle Schools

The purpose of middle school education is affected by the organization and articulation of K-12 schooling within a jurisdiction. Two common configurations of middle grade education are the middle school (5–8) and the junior high (7–9). The differences between a middle school and a junior high school are summarized in Figure 5–1.

	Middle School	Junior High School
Grades	5-8	7-9
Instructional Focus	Student-centered	Teacher-centered
Curriculum Source(s)	Affective and content driven	Content driven
Structure	Interdisciplinary teams	Departmentalized
Course Organization	Basics/exploratory courses	Required and elective courses
Purpose	Transition school and advisory program	Preparation for high school

Figure 5–1: Differences between middle school and junior high school (Adapted from Roth, 1991).

The recommendations for the effective education of young adolescents point toward the middle school model as an appropriate way to meet the developmental needs of students at this age. Bame (1986) summarized the purposes of middle school from the perspective of the young adolescent's development and suggested six functions that may provide practitioners with a framework for organizing middle school curriculum:

- *Socialization* - Preparation of students for entry and effective participation in society.

- *Articulation* - Recognition of the transitional nature of middle school and sequencing content and teaching methods to allow for the gradual movement of the student from the elementary (self-contained classroom) model to the high school (departmentalization) model of organization.

- *Differentiation* - Flexibility and adaptability in curriculum content and teaching methods that account for the diverse needs and abilities of learners at this age level.

- *Guidance* - Ways to assist students in making decisions concerning career choices and as well as decisions about personal and social actions.

- *Exploration* - A broad and general education designed to help students discover their potential vocational and avocational interests.

- *Integration* - Curriculum characterized by a wide range of interdisciplinary activities (pp. 72–73).

RECOMMENDED INSTRUCTIONAL METHODS

The foregoing discussion focuses on the purpose and organization of middle schools. The American Association for the Advancement of Science (1989) suggested some principles to guide instructional practice. They listed the following research-based characteristics of effective instruction in science, mathematics, and technology:

- Start where the students are with events and questions that are interesting and familiar to students and not outside their range of experience and understanding.

- Use active teaching methods that provide numerous opportunities for such activities as collecting, sorting, cataloging, observing, note taking, sketching, interviewing, surveying, measuring, and computing.

- Focus on the collection and use of evidence and allow students to decide relevance and meaning of the evidence gathered.

- Provide a historical context for scientific ideas.

- Insist on clear oral and written communication.

- Frequently use group activities to reinforce the importance of teamwork.

- Do not separate knowing from finding out.

- Focus on understanding rather than memorization of technical vocabulary (pp. 147–149).

Implementation of these instructional elements places more emphasis on understanding principles and concepts, and learning to learn in a way that is consistent with cognitive learning theory (Phye & Andre, 1986). In keeping with these ideas, Savage and Sterry (1990) suggested three instructional approaches that are appropriate for the study of technology: (1) the thematic or interdisciplinary approach, (2) the modular approach where reasonably self-contained activities relate to specific objectives, and (3) the problem-solving design brief or technology learning activity (TLA) approach where students are required to apply the technological method to solve a problem within given constraints.

CONSTRUCTION EDUCATION AT THE MIDDLE SCHOOL

How does construction technology fit in the middle school technology education program? Figure 5–2 depicts the Illinois scope and sequence for a K-adult curriculum model designed to prepare individuals for employment (Illinois State Board of Education, 1989). Consistent with other curriculum models of technology education, the middle school years emphasize technological and career explorations through practical problem-solving activities. Indeed, the International Technology Education Association (1985) noted that "problem solving, career orientation, and learning for tomorrow's adaptive environments are cornerstones of the technology education program at the middle school or junior high school" (p. 26).

Relationship To Technology Education

A common sequence in the middle school curriculum is to offer a one semester Introduction to Technology course in grades six or seven that allows for the broad exploration of the technologies in such areas as

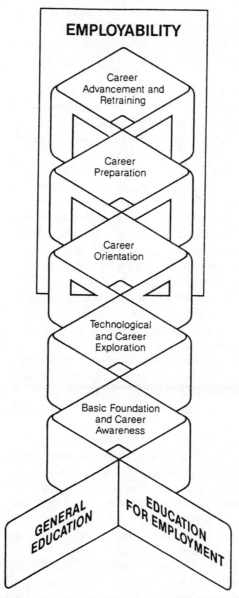

Figure 5–2: Scope and sequence for a K-adult Education for Employment Curriculum Model (Illinois State Board of Education, 1989).

communication, construction, manufacturing, transportation, and bio-related technology. This introductory course provides a foundation of concepts and principles for subsequent courses at the upper middle school grades that allow for a more in-depth exploration of each technological area. As can be seen in Figure 5–3, the construction unit is only one of 15 topics explored in a typical introduction to technology course.

Integrating Construction Technology. After the initial exploration of construction as part of an introduction to technology, construction may be studied as a separate system, as in the Jackson's Mill approach (Snyder & Hales, 1981), or included along with the study of manufacturing as part of the broader categorization of production systems.

Although the development of courses according to technology systems is widely practiced, Neden (1990) makes the point that most applications of technology do not fit into discrete categories such as manufacturing, construction, communication, and transportation. For example, Neden asked, "Where should laser technology be taught? By the communications teacher? The construction teacher? Manufacturing? By all?" (p. 26). Clearly, there are many opportunities for the study of laser applications in any of these areas of technology.

In response to this concern, a modular approach, independent of clusters, has been developed and implemented at middle schools in Pittsburg, KS and Delta, CO. The modular concept designed in Pittsburg, uses three "space stations" that accommodate 12 technology learning activities in the Intro-duction to Technology series in the sixth grade. Students then move on to the Explorations in Technology course, which includes 32 modules, in the seventh and eight grades. The focus of the program is an exploration of technologies and their applications across all career fields and not just industrial systems.

Working in pairs, students spend seven days completing the activities that comprise each module. Returning to the laser example, students actively investigate the operation of lasers through hands-on activities and resource material. After this initial investigation of concepts and operations, Larry Dunekak (telephone interview, January 28, 1993), an instructor at the Pittsburg Middle School, described using teacher-centered large-group settings to explore real-world applications of laser technologies in career fields as diverse as medicine and construction.

Figure 5–4 comprises a partial listing of technology modules used at the seventh and eighth grade level in the Pittsburg Middle School. While there is some overlap between the module titles and the units in a more traditional Introduction to Technology course (See Figure 5–3), the modules do not necessarily reflect an industrial base. Modules are designed to investigate

Computer technology	Mechanical and fluid systems	Technology and the future
Graphic communications	Materials and processes	Recycling materials
Electronic communications	Construction technology	Limited pollution
Electricity and electronics	Manufacturing systems	Solar energy
Transportation systems	Research and development	Alternate energy sources
Material handling systems	Product servicing	Careers in technology

Figure 5–3: *Typical units in an Introduction to Technology course (Adapted from the Illinois Plan for Industrial Technology Education).*

Applied physics	Electricity	Meteorology
Audio broadcasting	Electronics	Research and design
Computer graphics and animation	Energy, power, and mechanics	Robotics and automation
Computer problem solving	Engineering structures	Rocketry and space
Computer applications	Flight technology	Transportation
Desktop publishing	Graphic communications	

Figure 5–4:   Partial listing of modules used at the seventh and eighth grade level in the Pittsburg Middle School.

the technologies of meteorology as well as transportation; and construction is not specifically singled-out for study, but rather students explore a module entitled Engineering Structures.

Articulation Concerns. One of the major difficulties in dealing with the construction curriculum at the middle school level is the question of articulation. The differences between exploratory approaches in middle schools and the orientation curriculum at the early high school years are not well defined. Indeed, textbooks and curriculum materials aimed at those two markets often appear interchangeable. The once innovative bridge-building and testing activities may now be practiced at any level from elementary through high school, and in science as well as technology.

With the wide developmental difference in students between the sixth and ninth grade, and the existence of both middle and junior high schools, this lack of clear articulation between levels is perhaps understandable. Even the International Technology Education Association (1985) describes the goal of middle school or junior high school programs as "orientation and exploration . . . [designed to help students] make informed and meaningful educational and occupational choices" (p. 26). A challenge for practitioners may be to develop a more articulated sequence where competencies developed at the exploration level are related to, but different from, those capabilities developed at the orientation level.

Scope Of Exploratory Construction Technology Programs

The purpose of middle school technology education programs is to explore careers and technological processes. Therefore, the context for the selection of learning activities in an exploratory construction program may be the built world and beyond. The thematic activities for middle schools could be designed around: (1) the types of projects, (2) construction systems or project life cycles, (3) the management of the construction enterprise, and (4) the evolution of construction. The elements for the selection of exploratory activities in construction are presented in Figure 5–5.

Types of Projects. Construction ventures may be categorized by their intended use as either structures, civil projects, or transportation related projects. Structures may be further classified as using either rigid or non-rigid materials (Colelli, 1988). Rigid structures use elements such as frames, beams, and trusses. Examples of rigid structure construction range from traditional residential frame construction and commercial skyscrapers to geodesic domes and space frames. Non-rigid structures use materials such as cables and membranes. These structures may be inflatable or air

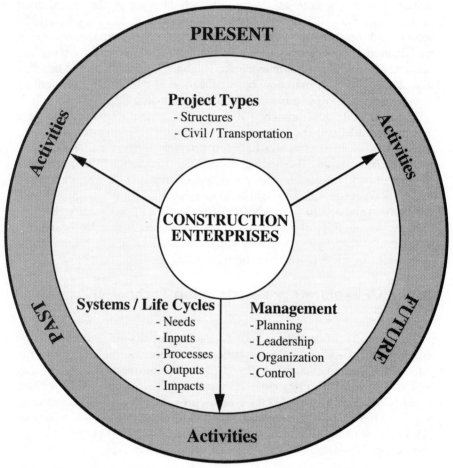

Figure 5–5:   *Elements for the selection of exploratory activities in con-*
struction.

supported like the Minneapolis Metrodome. An exploration of the construction enterprise should also consider civil and transportation projects like dams, towers, and pipelines, bridges, roads, highways, canals and locks, and runways.

Systems and Life Cycles. The systems approach of inputs, process, outcomes, and feedback, may also be combined with the idea of project life cycle as a way of organizing learning activities. The milestones of a construction project's life cycle include the inception or identification of the need; designing and engineering structures; planning and mobilizing for construction; assembly and installation on-site; service, maintenance and remodeling; and even the eventual demolition.

Management of the Construction Enterprise. The project life cycle could be overlaid with the enterprise aspects of managed construction systems. Student activities may then emphasize the functions involved in the management of architectural, engineering, or construction firms. This enterprise approach has a long history in technology education and industrial arts. Bame (1986) suggested the implementation of an enterprise model of organizing a business and then following through with actual or model project developments, making sure the base is broader than residential construction. Capitalizing on the interests of middle school children, the activities could also encompass the management of past and possible futures of construction as well as the present.

Evolution of Construction. Designing and modeling a space station, planning future cities, or investigating what it took to build castles in the middle ages are all technology-rich construction areas that have great potential for thematic activities. Johnson (1989) suggested that the design of a space station, complete with life-support systems "requires considerable imagination and forces students to search for advanced technologies in many fields" (p. 29). Recognizing the lure of space activities, Soman and Swernofsky (1993) used the planning of a space community as a synthesis activity. But activities requiring imagination and problem solving exist on terra firma as well as in space. Land-based activities might attempt to forecast and model cities of the future or address present community planning needs. Additionally, many of the concepts and principles of project organization and implementation that were used to build Roman roads or medieval castles have useful links to today's practices.

Implementation Strategies And Activities

Exploratory activities in construction technology can take many forms. Class organization may feature a mix of individual, small-group, and whole-class activities depending upon the objectives of the instruction. As Swernofsky (1989) noted, it is important to develop students' knowledge of construction materials and processes so that they have the tools to solve the problem effectively. For example, students may need to individually conduct library research or develop preliminary designs in order to contribute to the solution of a problem. Small groups may be formed to search for the solution to a problem or students may form teams to role-play the management of an enterprise. Whole-class safety demonstrations may be given as tools and equipment are introduced. Whole-class sessions may also be an effective way to debrief the concepts investigated through individual or small-group activity, reinforce interdisciplinary content, or explore career possibilities.

The following section highlights selected activities, approaches, and resource materials that may provide guidance in the development of exploratory construction activities. The activities usually employ a variety of student groupings and emphasize problem-solving, decision-making, and thematic or interdisciplinary instruction.

Construction and Thematic Exploration of Technology. At the Technology Festival of the International Technology Education Association Conference at Minneapolis in March 1992, Junior Branson of Southwest Junior High in Forest Lake, MN, displayed two Technology Learning Activities (TLA) for the exploration of construction technologies and related energy, transportation, and bio-tech concerns. The unit uses individual, small-group, and whole-class activities. The first activity was entitled Dream House. The assignment for each student was as follows:

> Imagine you were granted a wish. That wish is to draw a dream house with cost being no object and it can be located any place that you want to build it (space, underground, in the water, on an island, in the mountains, any place you would like it to be). Your house must include the following:

- The type of energy source you will be using;

- How you will get to school;

- Your food source;

- The location of Southwest Jr. High;

- Recreation facilities.

The second activity was a group project which built upon students' individual efforts in the first assignment. Entitled Interactive Community, students formed teams of four or five based on the commonalities of their original dream house design. That is, all of the students who located their dream home in space worked together as a group (to a maximum of four or five!). The problem for phase two was to:

Design a community using the dream house drawing as a base for the community. Items to be included in the community are:

- The type of energy system used in the city;

- Means of mass transportation;

- Food source and clothing (malls, stores, etc.);

- Government buildings;

- Southwest Jr. High;

- House of the person working on the project;

- Name of your town or community;

- Your community recreation.

To pull their ideas together and develop communication skills, each community planning team is required to give a presentation designed to persuade other class members that their community would be a wonderful place to live. Each presentation also includes a question-and-answer session where students must be prepared to justify their dream-community scenario.

Construction and Interdisciplinary Approaches. Illinois State Board of Education (1992b) has developed an interdisciplinary curriculum for Applied academics, Career exploration, and Technological literacy (ACT) at the middle school. The ACT Curriculum uses a systems approach and challenge problems to motivate students to thematically explore technology associated with an enterprise in their community. The ACT Curriculum was designed to reflect and help implement the recommendations for reform suggested by the Carnegie Council on Adolescent Development (1989).

The rationale for the thematic approach of the ACT Curriculum cites the works of Dewey (1938) and Whitehead (1952) and noted that most things in life are interdisciplinary and require synthesizing knowledge from a number of areas to solve realistic problems. Units related to construction are categorized in the Creating New Environments series and are designed to:

Introduce students to the technological concepts and careers associated with architecture and engineering. In addition, the unit provided a practical context for applying a variety of mathematics, science, and aesthetic concepts. The interdisciplinary topics addressed during the unit included things like measurement, basic geometry, mechanical forces, and design principles. (Appendix A, p. 1)

The three learning activities in the construction unit are titled: (1) Sticks, String, and Structures; (2) Spanning the distance, and (3) Thinking like Bucky Fuller. At the beginning of the unit, students establish their own architectural and engineering firms for the purpose of designing, building, and testing structures that address real problems. In the activity, Thinking like Bucky Fuller, students investigate geodesic domes while seeking to address the problem of providing temporary shelters for people made homeless through natural disasters.

Construction and Mathematics.   Komacek and Gorman (1992) reported how a technology teacher and a mathematics teacher combined to pilot test a geodesic-dome unit in a junior high school. The unit was introduced with slides of dome applications such as the Epcot Center and a discussion of the advantages and disadvantages of dome structures. A previously constructed paper model was then displayed to motivate the students and to demonstrate to the students that the structure really could be built! To construct the dome, each student was given a procedure sheet and a pre-cut kit.

In mathematics, students examined the concepts and methods required to calculate the size of the dome and length of the struts. The strength of the structure was also analyzed using an efficiency rating, that is the weight of the load causing failure divided by the weight of the dome. Analysis indicated that efficiency ratings of 100 were not uncommon. To further apply the structural concepts and challenge student thinking, "what if" questions were asked. Through example problems related to local structures, students investigated the application of dome engineering to bridges and calculated how many semi-trailer trucks could be supported at a efficiency ratio of 100.

Construction and History.   Activities may be designed to develop students' historical perspective of technology. The Nebraska Industrial Technology Education Exploratory Curriculum (no date) included a problem-solving assignment in which students "construct a model of an early Roman city including buildings, roads, aqueducts, and sewage systems" (p. 350). Similarly, an exploratory activity that focused on castle building in the Middle Ages would develop students' awareness of history while investigating construction problems, processes, and materials. Historical activities have tremendous opportunities for thematic cross-disciplinary instruction.

Construction and Computers.   The computer is arguably "the" tool of the late 20th century. Construction technology programs usually introduce students to computer-aided drafting (CAD) or perhaps some type of estimating software. Word-processing and desktop-publishing programs may also be used in enterprise activities to develop marketing materials or construction documentation. In addition to these computer applications, interactive computer simulations or games may be used to support the activities described above. For example, Castles by Interplay Productions (1991) is a program in which the intrepid builder must organize, schedule, and finance the construction of a castle amidst historical intrigue and the ever-present dangers of the marauding Celts!

Similarly, the SimCity (Maxis Software, 1992) series of software programs requires the novice community planner to manage various design and construction factors in simulated environments. Steve Murray, a middle school technology teacher in Leadwood, MO, uses SimCity as an extension to a community-planning-based activity called SuperCity 2000. Students work from a design brief to develop a model community for the year 2000. Students then transfer the information from their model into SimCity and run a simulation of their design. The simulation may result in a thriving metropolis with lots of people, cars, and tall buildings; or a rural community with a slower pace of life. In either case, as long as students can provide a place for people to live, work, and play while controlling traffic, pollution, overcrowding, crime, and taxes, the community will prosper. If not, people will leave the city creating slums, economic decay, and unrest. By using SimCity students see firsthand how their decisions impact a community's future.

SUMMARY

Construction may be explored in a variety of developmentally appropriate ways. Hands-on activities using thematic or interdisciplinary approaches is one of the key recommendations for matching instruction to the learning characteristics of middle school students. Students at this level have the capacity to deal with abstract and complex ideas but often need assistance in making connections between facts and systems. Thematic and interdisciplinary activities, such as those that link construction technologies to history or mathematics, help young adolescents make connections across subject areas.

The construction industry offers many career choices and has integral ties to most areas of technology. Therefore, a broad-based exploration of construction and related technologies can contribute to the growth and development of the young adolescent. Technology learning activities, such as Interactive Community or Thinking like Bucky Fuller, that emphasize problem solving and decision making provide opportunities for critical thinking in an atmosphere of diversity and choice. The opportunity for the exploration of personal interests within activities assists the young adolescent in discovering their likes and dislikes and, thereby, facilitates their search for personal identity.

Group simulations, such as the management of a construction or engineering enterprise, encourage interaction with peers and allow students of differing abilities to work together on a common project. Moreover, through role playing, values and team decision-making processes may be explored. These small-group activities provide supportive learning environments that encourage cooperative learning. Well-managed, small-group instruction provides opportunities for everyone to succeed, to help each other, and to learn to function as part of a team.

Finally, it is difficult to talk about technology and preparation for the future without talking about computers. Many opportunities exist for integrating computer applications into activities that explore construction. By using a range of software from simple drawing programs to game-type simulations, students develop competencies in applying the most commonly used contemporary tools.

Technology education at the middle school level has a significant role to play in helping youngsters choose a path towards a productive and rewarding future. The exploration of construction in ways consistent with the needs of young adolescents can help to clarify that choice.

REFERENCES

American Association for the Advancement of Science. (1989). *Science for all Americans; A project 2061 report on literacy goals in science, mathematics, and technology.* Washington, DC: Author.

Bame, E.A. (1986). Middle/junior high technology education. In R.E. Jones & J.R. Wright (Eds.), *Implementing technology education* (pp. 70–94). Mission Hills, CA: Glencoe.

Capelluti, J., & Stokes, D. (Eds.). (1991). *Middle level education: Programs, policies, and practices.* Reston, VA: National Association of Secondary School Principals.

Carnegie Council on Adolescent Development. (1989). *Turning points: Preparing American youth for the 21st century.* Washington, DC: Author.

Colelli, L.A. (1988). *Technology education: A primer.* Reston, VA: International Technology Education Association.

Dewey, J. (1938). *Experience and education.* New York: Macmillan.

Dunekack, L. (January 28, 1993). Telephone interview.

Epstein, J.L. (1990). *What matters in middle school grades—Grade span or practices?* Phi Delta Kappan, *71*(6), 438–444.

Gilstrap, R.L., Bierman, C., & McKnight, T.R. (1992). *Improving instruction in middle schools.* Bloomington, IN: Phi Delta Kappa.

Glatthorn, A.A., & Spencer, N.K. (1986). *Middle school/junior high principal's handbook: A practical guide for developing better schools.* Englewood, NJ: Prentice-Hall.

Huth, M.W. (1989). *Construction technology* (2nd. ed.). Albany, NY: Delmar.

Illinois State Board of Education. (1992a). *Illinois middle-level assessment: A look at the state-of-the-art in middle-grade practices.* Springfield, IL: Author.

Illinois State Board of Education. (1992b). *ACT: An interdisciplinary curriculum concept for applied academics, career exploration, and technological literacy.* Springfield, IL: Author.

Illinois State Board of Education (1989). *Industrial technology orientation guide.* Springfield, IL: Author.

Illinois State Board of Education. (1987). *Illinois plan for industrial technology education: An implementation guide.* Springfield, IL: Author.

International Technology Education Association. (1985). *Technology education: A perspective on implementation.* Reston, VA: Author.

Interplay Productions. (1991). *Castles.* [Computer program]. Santa Ana, CA: Author.

Johnson, J.R. (1989). *Technology: Report of the Project 2061 phase I technology panel.* Washington, DC: American Association for the Advancement of Science.

Kindred, L.W., Wolotkiewicz, R.J., Mickelson, J.M., & Coplein, L.E. (1981). *The middle school curriculum: A practitioner's handbook* (2nd ed.). Boston: Allyn and Bacon.

Komacek, S.A., & Gorman, P.S. (1992). Integrating mathematics with technology education: Geodesic domes. *The Technology Teacher, 51*(6), 17–21.

Lounsbury, J.H., & Clark, D.C. (1990). *Inside grade eight: From apathy to excitement.* Reston, VA: National Association of Secondary School Principals.

Maxis Software. (1992). *SimCity.* [Computer program]. Moraga, CA: Author.

Nebraska Industrial Technology Exploratory Curriculum. (n. d.). Lincoln, NB.

Neden, M. (1990). Delta County technology project. *The Technology Teacher, 51*(4), 25–29.

Orstein, A.C. (1992). *Secondary and middle school teaching methods.* New York: Harper-Collins.

Phye, G.D., & Andre, T. (1986). *Cognitive classroom learning: Understanding, thinking, and problem solving.* San Diego, CA: Academic.

Piaget, J. (1950). *The psychology of intelligence.* London: Routledge.

Rossi, K.D., & Stokes, D.A. (1991). Easing the transition from the middle level to the high school. In J. Capelluti & D. Stokes (Eds.), *Middle level education: Programs, policies, and practices* (pp. 19–23). Reston, VA: National Association of Secondary School Principals.

Roth, L. (1991). Middle level transition. In J. Capelluti & D. Stokes (Eds.), *Middle level education: Programs, policies, and practices* (pp. 42–48). Reston, VA: National Association of Secondary School Principals.

Savage, E., & Sterry, L. (1990). *A conceptual framework for technology education.* Reston, VA: International Technology Education Association.

Snyder, J.F., & Hales, J.A. (1981). *Jackson's Mill industrial arts curriculum theory.* Fairmont, WV: Fairmont State College.

Soman, S., & Swernofsky, N. (1993). *Experience technology: Communication, production, transportation, biotechnology.* Peoria, IL: Glencoe.

Swernofsky, N. (1989). *Making technology work.* Albany, NY: Delmar.

Toepfer, C.F. Jr. (1991). Organizing and grouping for instruction. In J. Capelluti & D. Stokes (Eds.), *Middle level education: Programs, policies, and practices* (pp. 24–29). Reston, VA: National Association of Secondary School Principals.

Whitehead, A.N. (1952). *The aims of education.* New York: Macmillan.

CHAPTER 6

Construction In High School Technology Education

Stanley A. Komacek, Ed.D.,
Department of Industry and Technology
California University of Pennsylvania, California, PA

This chapter examines curriculum and instruction for construction technology education at the high school level. The first section provides a review of the characteristics of high school students. The remaining sections answer three basic questions. Why study construction in high school? What should students study about construction in high school? How should construction technology be taught in the high school? The answers to these three questions form the basis for a program of study; namely a philosophy and rationale (why?), content and subject matter (what?), and instructional materials and strategies (how?).

CHARACTERISTICS OF HIGH SCHOOL STUDENTS

Any discussion of curriculum should begin with an analysis of the characteristics and needs of the learner. High school students, according to Newmann and Behar (1982) "seem to have grown quite close to what would be considered 'adult' capabilities in physical coordination, capacities for complex thought, social perspective taking, and other underlying abilities" (p. 27). Developmental psychologists describe high school age (approximately 15 to 17 years) as the middle-adolescence period. Middle adolescence is considered the most stressful stage in the transition from dependent child to self-sufficient adult (Harper & Marshall, 1991). Much of the stress during this period is associated with answering the question, "Where do I fit in?" According to Farrell (1990) "the American high school has been able to help large numbers of adolescents, perhaps even the majority, answer this question. But for many it has been ineffective" (p. 1).

This section presents two views of the characteristics of high school students. A set of "developmental tasks" for adolescents presented by Havighurst (1972) will be described first, followed by a model of "competing selfs" presented by Farrell (1990).

Havighurst's Developmental Tasks

Developmental tasks are the physical, social, and emotional milestones that occur as an individual matures. Havighurst (1972) described a developmental task as:

> a task which arises at about a certain period in life of the individual, successful achievement of which leads to his happiness and to success with later tasks, while failure leads to unhappiness in the individual, disapproval by the society, and difficulty with later tasks. (p. 2)

Havighurst (1972) identified eight developmental tasks of adolescence as follows:

1. **Achieving new and more mature relations with age-mates of both sexes.** Three subtasks demonstrate achievement of this developmental task: (a) learning to look at girls as women and boys as men, (b) learning to work with others while disregarding personal feelings, and (c) learning to lead without dominating. As these subtasks illustrate, social activities, not education, are a primary focus for high school students.

2. **Achieving a masculine or feminine social role.** This task involves learning and practicing a socially acceptable adult-masculine or feminine role. Masculine and feminine social roles have changed dramatically in the past few decades. Consequently, adolescents have more choices in selecting an acceptable role.

3. **Accepting one's physique and using the body effectively.** During adolescence, there are wide variations in the age at which physical maturation occurs. High school students are constantly comparing their physical appearance to their classmates. Any physical differences in body weight and form, facial features, hair style, or skin complexion can cause an individual to experience alienation and difficulty in achieving this task.

4. **Achieving emotional independence of parents and other adults.** To make the transition from childhood to adulthood, the adolescent must

become emotionally independent from parents and other adults while maintaining affection and respect. This task can result in a rebellious nature in high school students.

5. **Preparing for marriage and family life.** The idea of family life has also changed greatly in recent decades. The adolescent must develop an acceptable attitude toward marriage, family life, and child rearing.

6. **Preparing for a career.** According to Havighurst (1972) "vocational interests come to the fore" during adolescence (p. 43). However, most high schools direct their attention to preparing students for higher education, not a career. Many high school students cannot make the connection between school work and the world of work, which can make this task difficult to achieve.

7. **Acquiring a set of values and an ethical system.** Traditionally, the family and local community have helped adolescents achieve this task. Changes in family, community, and educational structures, as well as the myriad of technological advances, have challenged traditional systems of values and ethics in American society. The adolescent will acquire a set of values, but they may be different from those held by earlier generations.

8. **Desiring and achieving socially responsible behavior.** This task involves learning to live in and contribute to our democratic society. High school students should be encouraged to participate as responsible adults in the democratic decision-making process at local, regional, and national levels.

Farrell's Competing Selfs

Adolescents are trying to create their identity, to become adults in charge of their lives, to become persons who know who they are, to identify their particular sense of self. Farrell (1990) suggests that the self is "a system created in our interpersonal relationships" and to develop this sense of self "the adolescent must integrate a number of what might be competing selves" (p. 3). Farrell identified seven competing selves; self-in-family, sexual self, self-as-loyal friend, self-in-peer group, self-as-student, self-as-parent, and self-as-my-work.

Self-in-family. In early childhood, the family is the central focus for the development of the individual. Adolescence is a transition from childhood to adulthood. To make this transition, the adolescent must achieve emotional independence from parents and other adults. Breaking the parent-child bond can make the family, and its inherent value system, a source of irritation and annoyance for the adolescent. Adolescents struggling for autonomy tend to evaluate their actions increasingly in terms of internal standards, or those of respected friends and peers (Newmann & Behar, 1982). Discrepancies between family values and those of peers become a source of frustration for the adolescent, and other family members. Often, the struggle to break away emotionally from parents and adults can lead to rebellious behavior at home and at school.

Self-as-loyal-friend. The bonds between friends can be stronger than family bonds for high school students. Often students can confide in friends more easily than they can in parents and teachers. The self-as-loyal-friend is closely related to the self-in-peer-group.

Self-in-peer-group. Adolescence is a turbulent period characterized by a series of rapid and unpredictable changes primarily in physical, emotional, and social development. The onset of these changes can vary widely leading to feelings of alienation. The peer group gives the adolescent a sense of fitting-in and belonging. Peer approval has a very powerful influence over high school students. They often become "slaves" to the conventions of their group in important matters like clothing, hair styles, music, use of slang, and attitudes toward education. According to Farrell (1990) "the peer group often determines how an adolescent talks, what he eats, what he wears, with whom else he associates, and how hard he works in school" (p. 4). High school students often adopt the values and morals of their peer group, even when they conflict with those of the immediate family.

Self-as-student. From the perspective of the school, the self-as-student should be the adolescent's primary identity, despite the tremendous pull from the other selfs. The ideal high school student would take a full load of courses, participate in extracurricular activities after school, and spend several hours each night on homework. The ideal student would be totally involved in school and his/her world would revolve around education. Farrell (1990) suggests this image of the ideal student is unrealistic. Additionally, he contends the conflict between such societal expectations and the reality of competing selfs can cause stress for the high school student.

Intellectually, most high school students have reached the formal-operational period, characterized by the ability to develop and systematically test hypotheses. According to Lindgren & Suter (1985) the cognitive processes possible during this period "are frequently encountered in technological occupations, in scientific research, and in management of organizations" (p. 49). Generally, high school students have the intellectual capabilities to participate in the world of work. Unfortunately, most cannot see the connection between school work and the world of work beyond education.

Farrell's research suggests most students find school boring and irrelevant (1990). According to Newmann and Behar (1982) educators, parents, and government leaders concede that "the hard work demanded for high achievement in school has no immediate relevance to problems that adolescents perceive in daily life" (p. 26). Most high school students concentrate on social and emotional issues, not education.

The lack of relevance of schooling contributes to the drop-out rate. Nationally, about 25% of secondary school students drop out each year. In some large urban areas and certain rural districts, drop-out rates can exceed 50%. The societal costs of dropping out include increased crime rates, higher unemployment, increased reliance on welfare, and decreased basic literacy.

Self-as-my-work. Self-as-my-work varies for college-bound and non-college-bound high school students. College-bound students emphasize development of self-as-student and getting accepted into college. Their primary work during high school is self-as-student. Non-college-bound students, however, face greater pressures to develop their self-as-my-work.

To be successful in college or the work world, Farrell (1990) recommends that both college-bound and work-bound students need to develop a sense of industry; "the desire to bring productive situations to completion" (p. 101). Farrell criticizes education for not helping young people develop their sense of industry. He targets the awarding of partial credit and providing rewards for trying hard when important tasks are left incomplete. Such actions, he contends, make the teacher feel more humane and sensitive, but prevent students from developing their sense of industry. Students must come to believe they can complete what they start. They must see their completed works and feel good about them. Unless this happens, "school is toil but not meaningful work" (Farrell, 1990, p. 15).

Sexual self. High school age students know about sex. They are familiar with contraception and many have seen the consequences of AIDS. At times, suggests Farrell (1990), "the sexual self is the primary self of adolescence, which can overwhelm all other selves" (p. 4). The self-as-student and self-as-my-work cannot compete with the curiosity and accompanying fantasies created by the sexual self. Physically, high school students are reaching full maturity. Socially and emotionally, they are constantly experimenting to develop acceptable adult-social roles and the high school becomes their "social laboratory" (Havighurst, 1972, p. 45). Farrell's research suggested that some students view sex as a rite of passage into adulthood (1990). According to Zelnik and Shah (1983) the average age at which adolescents have their first sexual experience is age 16 for females and age 15 for males. They also reported that 50% of teenage females and 70% of teenage males have engaged in premarital sex. Unfortunately, even when they know about contraception, sexual experimentation can lead to a premature development of self-as-parent.

Self-as-parent. According to Kleinfeld and Young (1989), almost 90% of sexually active teenagers do not become pregnant. The 10% that do, total an estimated one million teenage pregnancies per year. According to Farrell (1990) over 90% of high school age parents keep their babies. The responsibilities of being a parent further reduces the emphasis placed on the self-as-student. Boyer (1983) reported that "pregnancy and marriage are the main reasons for dropping-out given by white women, and are cited even more frequently by nonwhite women" (p. 244). According to Farrell (1990) 40% of high school drop outs cite pregnancy as the reason.

Substance Abuse And Suicide

A discussion of the characteristics of high school students would not be complete without addressing the issues of substance abuse and suicide. The rate of substance abuse among American youth is the highest in the industrialized world (Botvin, Baker, Dusenbury, Tortu, & Botvin, 1990). In recent years, drug and alcohol use among middle-class adolescents has been decreasing gradually. Rhodes and Jason (1990) reported that marijuana and cocaine use among middle-class high school seniors decreased 35% and 49%, respectively, during the past nine years. The "Just say 'no'!" and "this egg is your brains on drugs" public service announcements seem to be making a positive impact. For adolescents from lower socioeconomic and minority groups, however, drug and alcohol use are increasing (Rhodes & Jason, 1990). For many adolescents, experimenting with drugs and alcohol is a rite of passage to adulthood, similar to sexual experimentation. The National Council on Alcoholism estimates three million teenagers are problem drinkers and nearly 60% of all high school seniors experiment with illicit drugs (Johnson, Pentz, Weber, Dwyer, Baer, MacKinnon, Hansen, & Flay, 1990). Peer pressure, family problems, and stress due to educational expectations are three factors significantly related to substance abuse. The effects of drug and alcohol use among adolescents include "mood changes, impaired judgement and motor function, decreased attention span, memory loss, [and] poor school performance" (Johnson, Pentz, Weber, Dwyer, Baer, MacKinnon, Hansen, & Flay, 1990).

Suicide is the second leading cause of death among adolescents. Between 1960 and 1980, the teenage suicide rate rose 287% (Lester, 1991). Estimates for high school students range from 8% to 11% for suicide attempts (National Center for Health Statistics, 1989). In 1987, there were 5175 suicides among people younger than age 25. Researchers suggest that stressful events, such as arguments with parents or a boyfriend/girlfriend and problems with school, trigger teenage suicides, which are related to major depression, substance abuse, and a family history of suicide (Shafer, Garland, Gould, Fisher, & Trautman, 1988; Hoborman & Garfinkel, 1988).

WHY STUDY CONSTRUCTION IN THE HIGH SCHOOL?

There are many reasons why high school students should study construction. In this chapter, three reasons will be presented: (1) construction is a basic technology, (2) construction has a positive impact on our national economy, and (3) construction curricula provides excellent opportunities for the integration of technology, mathematics, and science.

Construction Is A Basic System Of Technology

Constructing shelters and other structures is a natural part of being human. In prehistoric times people sought shelter from the wind, rain, snow, heat, cold, and other natural elements in existing structures in the natural environment (e.g., in caves, under trees, and between rocks). As technology developed, humans met their need for secure, safe, comfortable shelter by designing and building structures in a variety of forms (e.g., tents, huts, yurts, igloos, and teepees) that provided artificial environments. Throughout history humans have always attempted to improve the materials, processes, techniques, and designs of the construction system. A few notable recent innovations in construction technology include the following:

- *Chunnel* - the 31-mile-long tunnel being constructed 300 feet under the English Channel to connect the United Kingdom and France;

- *Plastic House* - the futuristic home experiment conducted by General Electric that uses plastics as the primary building material;

- *Smart House* - the futuristic home experiment conducted by the National Association of Home Builders Research Foundation that uses microprocessors to control household systems such as heating and cooling, lighting, ventilation, security, fire alarms, and food preparation systems;

- *Fabric enclosure* - roofs on sports stadiums and other large, open-air areas made from lightweight teflon-coated plastic fabrics supported by pneumatic pressure;

- *High-rise structures* - skyscrapers towering over one-quarter mile into the air with tuned dynamic dampening systems that help these magnificent structures absorb tremendous wind loads; and

- *Biosphere* - a space-frame-type structure that encloses a self-contained miniature earth environment where scientists hope to learn more about the effects of pollution and the ability of humans to thrive in a closed environment.

Today, the built environment may seem more natural to many humans than nature itself. How many of us could survive even one night without the comforts provided by our technological shelters? Even when we "rough it" and "get back to nature," we shelter ourselves in futuristic-looking dome-shaped tents made from fiber-composite poles and lightweight plastic-fabric roofs and floors that are water-repellent, tear-proof, fire-resistant, insect-proof, and somewhat inexpensive. Humans depend on construction for basic shelter.

The need for shelter has evolved beyond basic needs to provide the complete artificial environment—the city. The "concrete jungle" includes towering steel and glass skyscrapers, sports arenas with plastic pneumatic roofs, extensive networks of underground subway tunnels and pipelines for water, electricity, natural gas, and sewage disposal, bridges, highway systems, airports, and the sprawling, massive, multi-level, meet-your-every-need shopping mall! No doubt about it, construction is a basic part of human existence in our technological society.

Technology educators have recognized the importance of construction and identified it, with communication, manufacturing, and transportation, as a basic determiner of technological development in human societies throughout history. The *Jackson's Mill Industrial Arts Curriculum Theory* (Snyder & Hales, 1981) provided a rationale for studying construction and the other three basic technological systems:

> Each of these systems and their subsystems has been in existence at some level of development throughout the history of civilization. Each has a central theme, is universal in all societies, has unique questions and problems, and contributes in some way to the survival and potential of human beings. (p. 3)

As a basic characteristic of our technological, democratic society it is important for all citizens to understand construction systems in order for them to participate in the democratic decision-making process. According to Maley (1985):

> The purpose of any education . . . must be linked or integrally tied with the functions of the citizen in this democratic society, since a democracy depends upon an informed citizenry. The role of technology, as well as the obvious need for technological "fixes" to societal problems in the future, will demand levels of technological understanding greater than ever before if the citizenry are to share in the decision-making process. (p. 5)

Our technological society needs citizens who understand the behavior of construction systems and can make informed decisions about their implementation. High school students will soon be working, voting citizens who may have to answer questions similar to the following:

- Should I build a new house, buy and renovate an older one, or rent?

- What is the payback period for insulating a home, or installing new doors and windows, or sealing and caulking possible leaks?

- Will the local community need, and can it afford, a new high school in the next decade because of increasing population in the area and increased enrollments in the elementary schools?

- How will a proposed 15-acre paved parking lot surrounding a new shopping mall affect runoff and drainage systems in the local community?

- Should federal, state, and local governments direct more funds to the improvement and maintenance of the transportation infrastructure, such as highways, bridges, tunnels, airports, etc., to enhance economic development?

- Are the benefits gained by damming a river (e.g., downstream flood control, hydroelectric generation, and improved aquatic recreational opportunities) worth the negative consequences of flooding homes and business in several communities under the proposed lake?

- What positive and negative economic, environmental, aesthetic, and human health impacts will result if local zoning codes are altered to permit the development of a new industrial park?

- What alternatives are available to provide affordable public housing for the homeless?

• Are there career opportunities for me in construction and its related fields?

In conclusion, construction in high school technology education should help students understand the behavior of construction systems by providing opportunities for analyzing its relationship to other systems of technology, societal systems, and biological systems such as the natural environment. Also, it should prepare students to make informed education and career decisions and to participate fully and intelligently in democratic decisions related to the implementation and potential impacts of construction systems.

The Economic Impact Of Construction

Understanding the behavior of construction systems involves recognizing its economic impact and worth. Construction contributes greatly to our economy. One example is the analysis of housing starts (the number of new home construction projects begun). Each month, the financial industry reviews this statistic as a barometer of the economic health of our nation. One reason for this monthly review is the tremendous amount of money required to buy materials and pay workers to build a new home. A recent report by the Business Council for Effective Literacy (BCEL) on construction spending during 1990 revealed "a total of $434.5 billion was spent on new construction" (1991, p. 9). Residential construction (i.e., houses, apartments, other homes) accounted for the largest amount, $187 billion (43%). The other new construction spending included $137.5 billion (31.6%) for nonresidential construction (i.e., industrial and commercial buildings) and $110 billion (25.3%) for pubic works projects (i.e., highways, bridges, water and sewer systems, schools, hospitals).

Another reason for the monthly review of housing starts is the anticipation of additional financial transactions. Consumer purchases of new furniture, large and small household appliances, home entertainment systems, and other necessary furnishings often follow spending on new home construction. Also, local governments receive additional revenues through the payment of school and real-estate taxes. The BCEL reported that "each dollar spent on new construction in the U.S. generates more than $3.60 in economic activity among retailers, utilities, transportation, computer services, and the like" (1991, p. 9).

Another example of the economic impact of construction is job creation. The BCEL reported that "each $1 million spent on new construction creates nearly 47,000 jobs" (1991, p. 9). The jobs created are not only in construction but other fields as well due to the positive impact of construction spending on other economic activities. Henak (1991) reported that construction and its related support industries provide 20% of all jobs in the U.S. workforce. According to 1991 data from the U.S. Bureau of Labor Statistics, the construction industry alone directly employed 4.8 million workers. That is down from a high employment number of 5.2 million workers in 1990, but recent estimates project growth to 5.8 million construction workers by the year 2000 (cited in BCEL, 1991).

As in other fields, the level of intellectual skills required by construction workers is increasing. According to the BCEL (1991):

> The building and construction industry, like every other, has become more complex and technical. In the past, this was a field where a strong young man without much education could always get some kind of work, but with the introduction of labor saving devices . . . the need for unskilled labor has decreased substantially. Higher basic skills are now required at every level and the old "I can figure it out" approach will not suffice. (p. 1)

The BCEL (1991) also suggested that "2/3 of workers in construction firms today have skills below those needed on their jobs" (p. 9). Coinciding with the call for increased basic skills, several reports have addressed the possibility of labor shortages in the construction industry (Home Builders Institute, 1991; Gasperow, 1988; Henry, 1989), which could mean increased attention to construction education programs.

In summary, construction in high school technology education should address the positive economic impact of construction. Also, it should inform young people about available career opportunities in construction-related fields and provide basic intellectual and technical skill development.

Subject Matter Integration

Recently, numerous reports have called attention to the need for subject matter integration; particularly of technology, mathematics, and science (e.g., Technology Education Advisory Council, 1988; American Association for the Advancement of Science, 1989; U.S. Department of Labor, 1991; U.S. Department of Labor, 1992). *What Work Requires of Schools* (U.S. Department of Labor, 1991), prepared by the Secretary's Commission on Achieving Necessary Skills (SCANS), identified technology as one of five basic competencies as "essential preparation for all students, both those going directly to work and those planning further education" (p. vi). The list included:

- *Resources* - allocating time, money, materials, space, and staff;

- *Interpersonal Skills* - working on teams, teaching others, serving customers, leading, negotiating, and working well with people from culturally diverse backgrounds;

- *Information* - acquiring and evaluating data, organizing and maintaining files, interpreting and communication, and using computers to process information;

- *Systems* - understanding social, organizational, and technological systems, monitoring and correcting performance, and designing or improving systems;

- *Technology* - selecting equipment and tools, applying technology to specific tasks, and maintaining and troubleshooting technologies. (p. vii)

A follow-up document, *SCANS in the Schools* (U.S. Department of Labor, 1992), suggested that the basic competencies, including technology, "can and should be integrated into each subject" (p. 1). The SCANS team justified their call for subject matter integration:

In the workplace, we do not use one skill at a time in isolation from other skills; effective performance requires many different skills used in combination. It stands to reason, then, that students benefit from working on tasks and problems that call on a range of skills. (p. 9)

Another report, *Project 2061: Science for All Americans* (1989), produced by the American Association for the Advancement of Science (AAAS), focused on "scientific literacy—which embraces science, mathematics, and technology" (p. 3). As an association supporting science, AAAS may think the ultimate goal is scientific literacy, but according to Irene Hayes, Manager of the Science Education Center at Battelle-Pacific Northwest Laboratories, "the kind of science that accomplishes most of the world's work is not science at all, but technology" (1992). Still, the subject matter integration ideas suggested in *Project 2061* portray technology education in a meaningful educational role.

As a high school subject, construction provides excellent opportunities to integrate various subjects, particularly mathematics and science, with technology studies in real-world contexts. There are many ways to integrate subject matter. For technology educators, the technological application should be the focus of student learning experiences with content and concepts from other subjects integrated to help students understand the behavior of construction systems.

Construction in high school technology education should integrate mathematics by allowing students to estimate and calculate materials and costs, plan construction time schedules, measure in fractional inches, area, volume, and degrees, and calculate loads and forces and structural efficiency ratings. Science should be integrated by allowing students to test the laws of physical science that control the stability, strength, and rigidity of structural elements and systems. Also, students can analyze the impacts of a proposed structure on local environmental conditions and investigate the application of solar energy, wind engineering, earthquake resistance considerations, and other scientific principles to structural design.

Basic communication skills can also be integrated in high school construction technology education. Students can write specifications, create sketches and drawings, read technical and reference manuals, illustrate details of construction management schedules with charts, graphs, and illustrations, and make presentations of work portfolios to their classmates and evaluation teams.

WHAT SHOULD BE TAUGHT?

This section will attempt to answer the question, "What should be taught about construction in high school technology education?" Answering this question involves specifying curriculum and content. However, before identifying specific content the role of high school technology education will be presented. Then, a review of construction curriculum guides will be presented to help identify content. Finally, a proposal for future construction curriculum development for high school technology education will be presented.

Construction In High School Technology Education

In *Technology: A National Imperative* (1988) the Technology Education Advisory Council of the International Technology Education Association (ITEA) provided an excellent overview of the role and purpose of technology education. The Council described technology as "a discipline whose practitioners are involved in the *systematic* study of the *creation, utilization,* and *behavior* of *adaptive systems* leading to the goal of technological literacy" (p. 10). One essential element identified by the Council of curricula designed for technological literacy was "studying the behavior of various technological systems" (p. 10). The Council also suggested that the behavior of technology "cannot be viewed apart from its interactions with society and the values of society" (p. 5). Figure 6–1 presents a model proposed by the Council to illustrate the interaction between technology, society, and values.

The Council described high school technology education as "designed to create outcomes related to scientific principles, engineering concepts and technological systems" (p. 19). They identified six student-learning outcomes

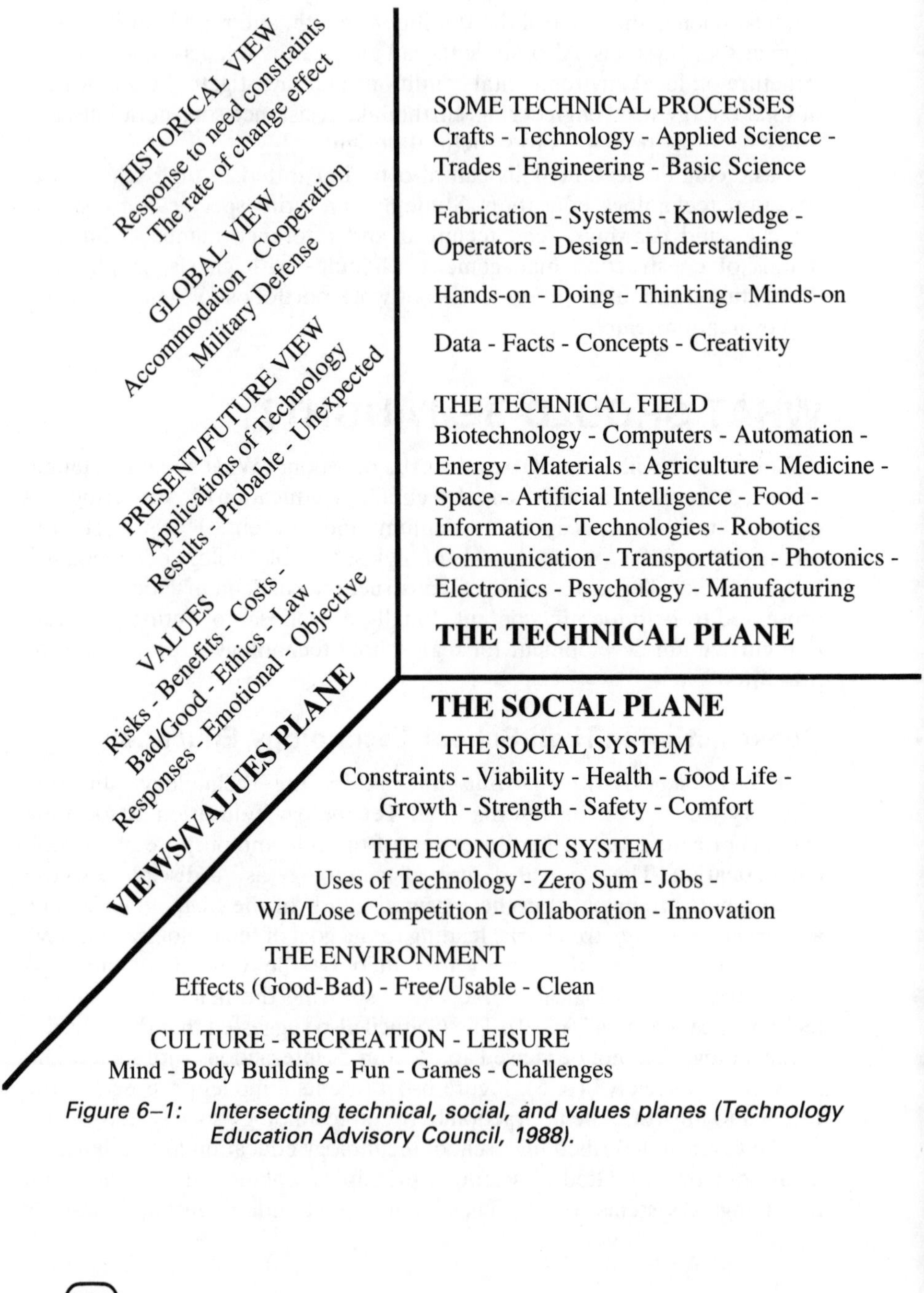

SOME TECHNICAL PROCESSES
Crafts - Technology - Applied Science -
Trades - Engineering - Basic Science

Fabrication - Systems - Knowledge -
Operators - Design - Understanding

Hands-on - Doing - Thinking - Minds-on

Data - Facts - Concepts - Creativity

THE TECHNICAL FIELD
Biotechnology - Computers - Automation -
Energy - Materials - Agriculture - Medicine -
Space - Artificial Intelligence - Food -
Information - Technologies - Robotics
Communication - Transportation - Photonics -
Electronics - Psychology - Manufacturing

THE TECHNICAL PLANE

THE SOCIAL PLANE

THE SOCIAL SYSTEM
Constraints - Viability - Health - Good Life -
Growth - Strength - Safety - Comfort

THE ECONOMIC SYSTEM
Uses of Technology - Zero Sum - Jobs -
Win/Lose Competition - Collaboration - Innovation

THE ENVIRONMENT
Effects (Good-Bad) - Free/Usable - Clean

CULTURE - RECREATION - LEISURE
Mind - Body Building - Fun - Games - Challenges

HISTORICAL VIEW
Response to need constraints
The rate of change effect

GLOBAL VIEW
Accommodation - Cooperation
Military Defense

PRESENT/FUTURE VIEW
Applications of Technology
Results - Probable - Unexpected

VALUES
Risks - Benefits - Costs -
Bad/Good - Ethics - Law
Responses - Emotional - Objective

VIEWS/VALUES PLANE

*Figure 6–1:   Intersecting technical, social, and values planes (Technology
Education Advisory Council, 1988).*

to further detail the purposes of high school technology education. According to the Council, high school students who study technology will:

- Experience the practical application of basic scientific and mathematics principles.

- Make decisions about postsecondary technology careers, engineering programs, or service-related fields.

- Make decisions about advanced technical education programs.

- Gain an in-depth understanding and appreciation for technology in our society and culture.

- Develop basic skills in the proper use of tools, machines, materials, and processes.

- Solve problems involving the tools, machines, materials, processes, products, and services of industry and technology (1988, p. 19).

The West Virginia scope and sequence model (Figure 6–2) illustrates the typical organization for technology education in elementary, middle, and high schools. According to the model, elementary school students should develop an awareness of technology through integrated studies, middle school students should engage in an exploration of the four basic technological systems, and high school students should analyze technologies in-depth. Many states, including West Virginia, do not require technology studies at the middle school level. The West Virginia model provides an opportunity for high school students to explore technology with an Introduction to Technological Systems course. As shown in the model, four systems courses are typically recommended, but several states combine construction and manufacturing into production. Advanced Technological Studies is not one course, but a category of additional courses that could be developed to follow each system course. The Advanced Technological Studies category gives schools the option to develop specialized in-depth courses.

Construction Technology Education Curricula

Curriculum guides often dictate or recommend what should be taught. In the area of construction technology education, most curriculum guides seem to follow the direction set by the Industrial Arts Curriculum Project (IACP) of the 1960s. According to Blankenbaker (1982), IACP defined construction as:

The practices mankind utilizes to build structures or other constructed works on a site. This definition includes the management practices

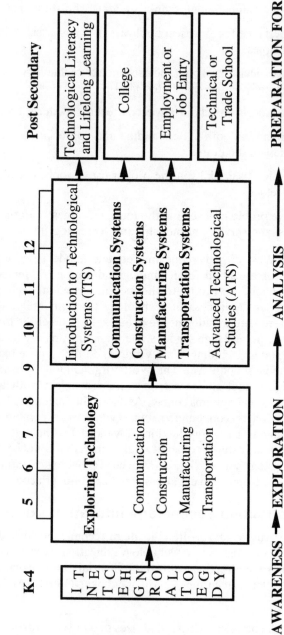

Figure 6–2: *Technology education scope and sequence model (West Virginia Department of Education, 1987).*

necessary to initiate, design, engineer, finance, erect, transfer, and service constructed works. Construction technology encompasses the production practices necessary to clear the site, build the structure, landscape the site, and service the finished project. This definition also includes personnel practices which enable people to work together to produce the desired construction project. (p. 31)

The IACP's emphasis on management, production, and personnel practices has served as a model for the development of many construction curriculum guides. One example is Exploring Construction Technology (Wood, 1987) by the Mid-America Vocational Curriculum Consortium (MAVCC). The MAVCC guide includes management practices (Units III and IV), production practices (Units IV, V, and VI), and personnel practices (Unit I) in its content outline Figure 6–3. Another example is Wisconsin's construction activity guide, which organizes instruction into three units; planning processes, production processes, and managerial processes (Kitzmann, 1988).

The educational goals of Exploring Construction Technology (Wood, 1987) and many other construction curriculum guides, appear to be understanding the construction industry, not an in-depth understanding of the behavior of construction systems as recommended by the ITEA's Technology Education Advisory Council. As Johnson, a member of the ITEA's Technology Education Advisory Council, suggested "most technology education programs . . . still focus on the technical plane" (1992, p. 4). The key production practices traditionally found in construction curriculum guides include designing and planning, site selection and preparation, excavating, setting foundations, building a superstructure, installing utilities, enclosing and finishing, landscaping, and transferring and servicing. Often, a large central project, such as a storage shed, is recommended and used to organize instruction and student learning experiences. With the focus on project construction, much of the instruction is on the technical plane with an emphasis on processing materials and building a structure. The focus on industry and the technical plane usually gives the impression of vocational education (trade training) to construction technology courses, which can cause confusion for students, parents, administrators, and teachers.

Construction: Implementation And Enrollments

Exploratory construction courses have not been widely implemented. According to Henak (1991) about one-third of states have implemented such courses. At the classroom level, construction courses are not very popular. According to the most recent version of the annual survey of the

UNIT I: Overview of Construction
Types of Construction
Resources Used in Construction
Steps in Construction
Construction Workers
Education and Training
Personnel Practices
Labor/Management
Construction and the Future

UNIT II: Construction Safety
Laboratory and Site Safety
Personal Safety
Housekeeping Practices
Ladder Use
Safety Rules - Power Tools

UNIT III: Design and Planning
Design and Engineering Process
Types of Public Construction Projects
Building Codes
Building Standards and Material Specs.
Soil Tests
Legal Land Descriptions
Working Drawings
Cost Estimating
Securing Building Permits
Types of Bridges

UNIT IV: Construction Preparation
Surveying and Equipment
Contracts and Duties
Securing a Contract
Scheduling Construction
Preparing to Build
Site Clearing
Earthmoving Equipment
Excavating

UNIT V: Construction Processes
Foundation Types
Foundation Materials
Setting Foundations
Superstructure Frames
    (Steel, Concrete, Wood)
Erecting a Structure
    (Floors, Walls, Roof)
Utility Systems
Insulating Structures
Enclosing the Exterior
Enclosing the Interior
Finishing the Project

UNIT VI: Project Completion
Landscaping
Transferring the Project
Final Inspection
Closing the Contract
Servicing the Project

field conducted by Dugger and others (1992) only about 14.5% of responding teachers reported course offerings in construction technology. The focus on construction industry practices may be one reason construction courses are not very popular. According to a study conducted for the Construction Industry Workforce Foundation (CIWF), young people have a poor image of construction (cited in Home Builders Institute, 1991). According to the Home Builders Institute, the CIWF study, in which junior and senior high school students were interviewed about construction, suggested that:

> Females feel . . . excluded by the construction industry.
> Young people have negative perceptions about women in the industry . . .
> Both male and female students say they are surprised when they see a female construction worker who is attractive. . . .
> Construction work is associated with dirt, sweat, and a rough demeanor.
> The construction worker lacks prestige and respectability.
> Construction work is boring, tiring, and stressful. (1991, p. 5)

The Home Builders Institute expressed concern that these negative perceptions about construction "may hamper the industry's ability to compete with other industries in recruiting new entrants to the workforce" (1991, p. 5). Similar perceptions of construction technology education courses may be hampering the efforts of technology teachers to compete with other courses in recruiting new students for their classes. Also, technology courses, particularly at the high school level, have traditionally been male dominated, which excludes the female portion of the population.

The apparent vocational orientation of construction technology education courses also may be hampering teachers' student-recruitment efforts. Technology educators profess to be a part of general education, but perceptions of students, parents, and school administrators may be more important in determining course enrollments. According to Boyer (1983) nationwide "about 11% of all high school students concentrate in vocational education—taking six or more such courses [and] another 18% take three vocational courses" (p. 120). If students perceive construction technology education courses as vocational education, only about 11% to 18% of the student body can be expected to enroll. Boyer (1983) also reported that in 1982 about 60% of the 3 million public high school graduates were college-bound. If students perceive construction technology education courses as inappropriate for college-prep majors, over half the student body may be excluded from enrolling.

A Role For Industry-Based Vocationally-Oriented Construction Courses

Industry- or industrial technology-based construction courses with a vocational orientation can play important roles in education. For example, Farrell (1990), who studied high school students at risk for dropping out, supported such a course. According to Farrell, at-risk students need to learn in a real-world context. He cited the example of an alternative course he taught to the Gang of Four (four at-risk male students) in which they began construction of an energy-efficient home. Over the course of an academic year, Farrell helped the students receive credit in economics (they budgeted finances when purchasing tools and materials, calculated labor costs, etc.); composition (they kept daily journals of building progress and wrote reports of what they learned); environmental studies (they researched frost-line depth, applied solar-design principles, studied the effects of runoff, conducted percolation tests, etc.); social studies (they researched building codes and secured permits, etc.); and mathematics (they calculated beam spans using algebra, laid out square rectangles using geometry, etc.). According to Farrell, the key for these at-risk students was the real-world context of their education:

> It made sense because we were in a real-world environment. They were learning to use the tools of their technology, from hammers to libraries, and they could see a direct relation between those tools and their lives. We saw the results of our labor each day and viewed our finished products. We had real-world accomplishments. . . . I could have taught a course in construction. . . . A course in construction might teach someone to *conceptualize* building houses, but the Gang of Four needed to *build* one. (1990, p. 159)

Dating back to industrial arts, teachers in our field have always played a significant role in helping educate at-risk students. In the future, high school construction technology teachers may continue to contribute in this area that demands so much attention. However, with increased emphasis on preparation for college, it seems that high school technology education needs new and different types of construction curricula. New construction courses are needed that attract female students, change the negative perceptions created by a focus on the construction industry, reduce the vocational orientation, and attract college-bound students. Very few examples of such curricula are available.

Indiana's Construction Curriculum. The State of Indiana has a well-developed set of comprehensive construction curricula. The Indiana Industrial Technology Education Curriculum (IITEC) includes guides for four construction courses—Introduction to Construction Technology (IITEC, 1992d), Construction Planning and Design (IITEC, 1992c), Constructing Structures (IITEC, 1992b), and Community Planning (IITEC, 1992a). At least two of the courses are unique for technology education. They break the mold of traditional construction curricula and hold promise for improving the image of technology education and attracting students who normally would not enroll in technology studies related to construction. One unique feature of the courses is the use of the project delivery process (PDP) model (Figure 6–4). According to Henak (1991), the PDP model illustrates the complete construction process, which "begins with an owner's needs and ends with his or her use of the structure" (p. 10). Each course is described below.

The modules in the Introduction to Construction Technology guide present a traditional view of construction organized to provide students with an overview of the entire PDP model. The construction industry, technical plane focus and vocational orientation established by IACP and supported in Exploring Construction Technology (Wood, 1987) are evident. Instruction is organized around a central construction project, a dog house called "Rover's Rad Pad."

Construction Planning and Design is a unique course. The introduction to this curriculum guide suggests that the course is "directed toward people who are interested in solving problems related to designing constructed projects" (IITEC, 1992c, p. I-4). Designing constructed projects is "architecture/engineering," which encompasses design sensitivity, knowledge of materials, an understanding of construction techniques, drawing skills, and an understanding of the PDP model (IITEC, 1992c, p. I-4). Instruction in the course modules is organized following the Designing/Planning component of the PDP model. Students work in teams and play the roles of architects, engineers, and managers as they design structures, prepare drawings, and write specifications. Much of the work is still in the technical plane, but the social plane and values plane are integrated.

Constructing Structures has a traditional orientation and focuses on the Constructing phase of the PDP. One module is dedicated to working with concrete and most of the other modules are organized around building a project (i.e., preparing a site, building a foundation, erecting a support structure, etc.). However, the course dedicates a large percentage of instruction time to two non-traditional modules, Constructing High-Rise Structures and Building Heavy Engineering Structures. Still, the developers seem unsure about student interest in the Constructing High-Rise Structures module. The introduction to the module states; "Hopefully, students

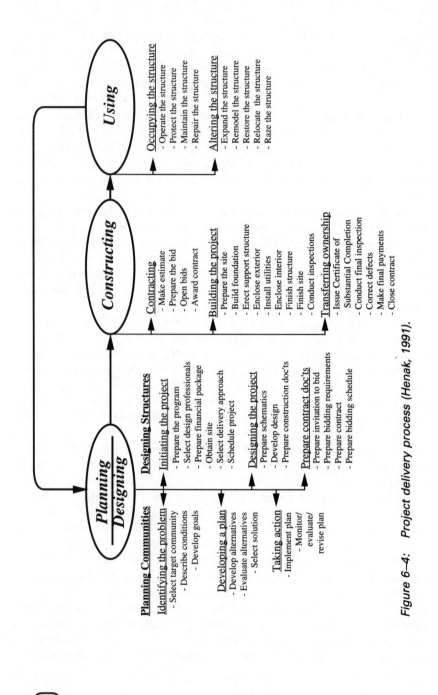

Figure 6–4: Project delivery process (Henak, 1991).

will be motivated through the study and analysis of prominent structures" (IITEC, 1992b, p. 8–1).

Community Planning (IITEC, 1992a) is another unique and interesting course. According to the IITEC the Community Planning course is:

> designed for people who are interested in solving physical, social, economic, and environmental problems of the built environment at a community scale. . . . The primary concern is with the physical elements; with the realization that there are social, economic, and political elements that shape the built environment. (p. I-4)

Content is organized around a model that engages students in three developmental phases; description, prediction, and action (Figure 6–5). According to Henak (1991) such a model facilitates the development of student learning activities "designed to provide plans of action for resolving current community concerns" (p. 11). The content outline includes module titles such as Identifying Community Concerns, Organizing the Planning Team, Analyzing Current Conditions, Determining Community Plans, and Implementing the Plan. This course holds promise for instruction that moves beyond the technical plane and addresses the interdependent and interactive behavior of construction systems, societal systems, and environmental systems. Such a course will not be easy to teach for an instructor whose primary focus has been traditional building construction. Extensive retraining or team teaching efforts with other disciplines will be required, but it should be worth the effort.

Future Construction Curriculum Development Efforts

The architecture and community planning courses from Indiana hold promise. They move beyond the technical plane, provide opportunities for the integration of other subject areas, shed the construction industry and vocational orientation perceptions, and may attract more female and college-bound students to construction technology studies. Another area that seems to hold promise is structural engineering.

Merritt described structural engineering as:

> the application of scientific principles to design of load-bearing walls, floors, roofs, foundations, and skeleton framing needed for the support of buildings and building components. . . . [Structural engineering] is the application of structural theory to ensure that buildings and other structures are built to support all loads and resist all constraining forces that may be reasonably expected to be imposed on them during their expected service life, without hazard to occupants or users and preferably without dangerous deformations, excessive sideways drift,

DEVELOPMENTAL PHASES	ELEMENTS OF A PHYSICAL PLAN							
	Environmental Protection	Land Use	Transportation	Housing	Economic Development	Public Facilities Utilities/Services	Fiscal Plan	Historic Preservation
DESCRIPTION PHASE: Diagnose Problem Set Goals								
PREDICTION PHASE: Develop Alternatives Evaluate Alternatives Selection Solution								
ACTION PHASE: Implement Plan Monitor/Evaluate/ Revise Plan								

Figure 6–5: Developmental phases used to organize content in Indiana's Community Planning construction course (Indiana Industrial Technology Education Curriculum, 1992a).

or annoying vibrations. In addition, good design requires that this objective be achieved economically. (1982a, p. 1–4; 1982b, p. 5–1)

Structural engineering involves considering spatial requirements, selecting materials, specifying structural elements, analyzing tension and compression forces, approximating dead loads, anticipating wind loads and earthquakes, predicting the effects of static and dynamic loads, considering deflection and deformation, and calculating the effects of fire (Tang & Chin, 1990). Some structures designed by structural engineers include "industrial buildings, tall buildings, bridges, thin-shell structures, arches, suspension roofs, tanks for liquid storage, bins and silos for granular materials, retaining walls, bulkheads, steel transmission towers and poles, chimneys, and buried conduits" (Gaylord & Gaylord, 1990, p. xv).

Many structural engineering-type student learning activities have been developed for technology education. Unfortunately, no comprehensive structural-engineering course for technology education is available. One resource technology educators could use as a starting point for the development of a structural-engineering course is Architecture and Engineering (Salvadori & Tempel, 1983). Figure 6–6 lists the suggested units of instruction in Architecture and Engineering.

This well-illustrated manual uses the hands-on/minds-on problem-solving method of instruction so familiar to technology teachers and includes numerous student-learning activities. The How-to-Use-This-Manual section explains how to sequence instruction and provides examples of preplanned instruction sequences for specific structures, such as bridges, skyscrapers,

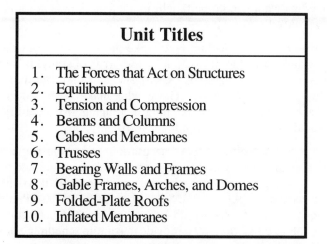

Unit Titles

1. The Forces that Act on Structures
2. Equilibrium
3. Tension and Compression
4. Beams and Columns
5. Cables and Membranes
6. Trusses
7. Bearing Walls and Frames
8. Gable Frames, Arches, and Domes
9. Folded-Plate Roofs
10. Inflated Membranes

Figure 6–6: Suggested units of instruction in Architecture and Engineering (Salvadori & Tempel, 1983).

playgrounds, inflated membranes, historical studies, literary and art studies, and others. Information on performing math calculations and graphing data is included, as well as suggestions for materials, supplies, and special equipment. An appendix includes a comprehensive glossary of defined terms and illustrated explanations of producing structural models using styrofoam.

Most of the instruction in a structural engineering course would still be in the technical plane. However, it would allow for the attainment of learning outcomes related to:

- understanding the behavior of construction technology systems;

- applying engineering concepts to the solution of technological problems;

- integrating science concepts and mathematics principles in practical situations; and

- helping students make decisions about education requirements and career opportunities in engineering and technology.

In addition, such a course would help construction in technology education shed the negative perceptions related to the construction industry and the limitations of a vocational orientation.

What Should Be Taught?   This section began by asking the question, "What should be taught about construction in high school technology education?" Several examples of construction content, courses, and curricula were presented to help technology educators answer the question. An easier question to answer may have been, "What *could* be taught?" Recommending what should be taught in a curriculum is not an easy task. According to Mullin (1991):

> To formulate a curriculum is to make judgements about the relative importance of different disciplines and areas of study. It is to declare that some subject matter is essential while other material is merely useful, but not necessary. Those who make such judgements lay themselves open to various charges, ranging from cultural bias to personal bias to narrow-mindedness. (p. 93)

Mullin suggests that some of these accusations may hold some degree of truth; and they are hard to refute, so many educators avoid taking a stand. Still, Mullin presents a challenge when he says "to refuse to articulate one's choice is cowardice" (1991, p. 93). So, four courses are described below to answer the question, "What could be taught about construction in high school technology education?" To provide maximum scheduling flexibility in most high schools, the courses could be non-sequential (i.e., no prerequisites) and designed for one semester of instruction.

Construction Systems, which would be similar to Introduction to Construction (IITEC, 1992d) and Exploring Construction Systems (Wood, 1987), would meet the needs of at-risk students and other students interested in developing construction vocation-related technical skills. A comprehensive overview of the PDP model would be presented in a real-world context using construction industry practices. Another option would be to present an introduction to construction by exploring major components of the three other courses described below.

An Architecture and Construction course, similar to Construction Planning and Design (IITEC, 1992c), would focus on the jobs of architects in the designing/planning phase of the PDP model. Student learning experiences would include analyzing an owner's needs, developing proposals, creating sketches and drawings, writing specifications, and preparing contract documents.

A third course, Structural Engineering has no technology education model. Architecture and Engineering (Salvadori & Tempel, 1983) could be used as a starting point. Students would use math, science, engineering, and technology principles to help them understand why buildings and other structures stand up.

Community Planning (IITEC, 1992a) is a model for a fourth course that would focus on a larger-than-structure scale. Students would address real problems in their local community related to construction systems and their interaction with society, the environment, politics, and economics. Such a course would move technology studies into the social and values planes proposed by the ITEA Technology Education Advisory Council.

HOW SHOULD IT BE TAUGHT?

In his research, Farrell (1990) discovered what many teachers already knew that most high school students find school boring. According to Farrell "what makes a class boring seems to be *how* it is taught rather than what is taught" (1990, p. 149). Much has been written about how technology education should be taught. The 37th yearbook of the Council on Technology Teacher Education, *Instructional Strategies for Technology Education* (Kemp & Schwaller, 1988), described ten different instructional strategies for technology education. Two monographs edited by Edmison and published by the ITEA in 1992, *Delivery Systems* and *Approaches*, described a total of 16 technology teaching strategies. This chapter will focus on one strategy; problem solving. Problem solving is not a strategy that works well for every technology learning activity, but it is the primary one for technology studies. A model for the organization of problem-solving-based student-learning activities in high school construction technology education will be presented. Sample learning activities will be used to illustrate the use of the model.

The Technological Problem-Solving Method

Many problem-solving models for technology education have been proposed (e.g., Page, Clarke, & Poole, 1982; Seymour, 1987; Hatch, 1988; Waetjen, 1989; New York State Education Department, 1990; Savage & Sterry, 1990; Komacek & Bolyard, 1992, in press). Some models are complicated, difficult to implement, have numerous steps, and can cause confusion between technology and science (Komacek & Bolyard, 1992, in press). One simple, easy-to-understand, and easy-to-implement problem-solving model is the technological problem-solving method (Figure 6–7). The four phases in the model—designing, producing, testing, and analyzing—provide one answer to the question, "How should construction technology be taught?" According to Komacek and Bolyard, "the four phases in the model ... describe what technologists do (and what technology students should do)" (1992, in press).

Following the technological problem-solving method, construction technology students would solve problems related to designing, producing (constructing), testing, and analyzing structures and other construction systems. Designing involves determining the nature of a human need or want, gathering information, considering alternatives, and recording ideas on paper (or on a computer screen) in sketches, drawings, and notes. Rough sketches, renderings, detail drawings, and comprehensive working drawings could be created. Producing involves constructing mock-ups and prototypes using tools, materials, and processes. Testing involves using a mock-up or prototype to check its function and operation. Analyzing involves evaluating a design idea, mock-up, prototype, or test result based on some criteria.

Relation Among the Phases. The illustration of the technological problem-solving method in Figure 6–7 represents the ideal sequence. Technological problems are not always solved in a clean, neat, step-by-step fashion. Usually, the problem solver must move back and forth among designing, producing, testing, and analyzing. As the arrows in Figure 6–7 suggest, the technological problem-solving method permits free movement among the four phases. For example, after initial designs are completed, the producing phase can begin. When production progresses smoothly, testing and analyzing can follow immediately. However, when design flaws are identified during production, work must return to the design phase. The same process holds true between and among all the phases in the method.

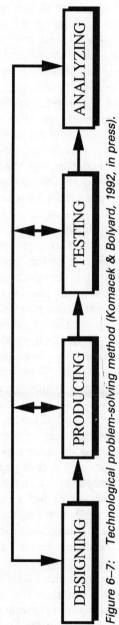

Figure 6–7: Technological problem-solving method (Komacek & Bolyard, 1992, in press).

DESIGNING → PRODUCING → TESTING → ANALYZING

Applying the Model to Construction. In a comprehensive construction course, students should be engaged in designing, producing, testing, and analyzing construction systems. Major problem-solving activities would engage students in all four phases and provide opportunities for free movement among the phases. All technological problems do not require designing, producing, testing, and analyzing to reach a solution. Some problems can be solved by completing one, two, or three of the phases. It follows that technological problem-solving activities could engage students in only one, two, or three phases. Four types of problem-solving activities are described below using the technological problem-solving method. Specific construction-learning activities are explained for each type.

Comprehensive learning activities engage students in all four phases of the model. Also, comprehensive activities provide students with sufficient time to move freely among the phases. In the real world, technologists often redesign, reproduce, retest and reanalyze their solutions. Freedom of movement among the phases allows students to study the behavior of construction systems in-depth, to a greater level of understanding. Of course, redesigning, reproducing, retesting, and reanalyzing takes time, which means less content coverage. Comprehensive activities that do not permit freedom of movement among the phases result in trial-and-error, not the preferred trial-and-adjustment. Students must be given opportunities to learn from their mistakes. If not, they may only learn they could not successfully solve the assigned problem. Often, the most enduring learning occurs when students adjust (correct) errors (mistakes). Comprehensive activities are usually of long duration. Three comprehensive high school construction technology education activities are described below.

House of Cards is an introductory activity that addresses dead-loads, live-loads, and structural-integrity concepts (Horton, Komacek, Thompson, & Wright, 1991). Although most comprehensive activities are of long duration, this one can be completed in two or three class periods. Student-design teams are given five index cards, one sheet of paper, 18 inches of tape, and 12 inches of string. Their challenge is to design a reinforcing technique for a house of cards made from two of the index cards (Figure 6–8). First, a mock-up structure is designed, produced, and tested. Then, based upon an analysis of the test results, the students redesign and produce a prototype for official testing and analysis. Assigned dollar values for the construction materials and the time used integrates economics. Engineering concepts and mathematics are integrated by calculating total cost, structural efficiency, and a safety factor. Evaluation is based on the students' understanding of key concepts and their contributions to the team effort. This is combined with an evaluation of the performance of the house of cards in three tests; structural-efficiency rating, maximum live load, and economic-efficiency

Live Load

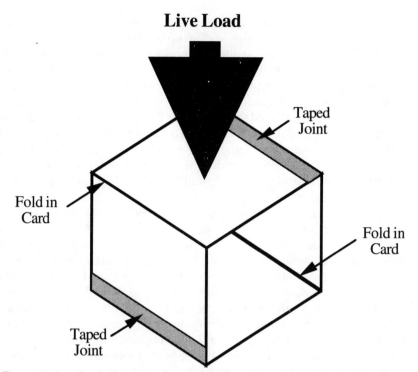

Figure 6–8: *Basic structure for house of cards activity (Horton, Komacek, Thompson, & Wright, 1991).*

rating. If redesigning and reproducing are not permitted, it is probably unfair to base a student's grade on the performance of the structure.

Bridge-building activities are popular in construction technology education. Bridge-engineering activities are not. Most technology education bridge-building activities involve designing, constructing, and destructively testing a model bridge. Students often get their ideas by looking at bridge designs. Usually, only one bridge is constructed for testing. Applying the technological problem-solving method, students should be given opportunities to repeat one or more of the phases, as needed. Also, designing should be based more on trying to predict the behavior of the structural elements in the bridge during loading.

Wynn's (1987) description of a bridge-engineering activity used with engineering students could serve as a model for construction technology teachers. His students designed an inverted bridge truss made from 3.175 mm square balsa members with sheet-balsa gussets. Before designing their bridge, students tested and analyzed the tension strength, lap-joint strength, butt-joint strength, and buckling resistance of the balsa members. Also, they analyzed the behavior of trusses and identified tension and compression members. Then, the students applied their understanding of the behavior of trusses and the results of the material tests to design their bridge. Completed bridges were destructively tested. Students did not reproduce their bridge after testing. However, destructive testing was video recorded. Students studied and analyzed the reasons for failure. The bridge could have been redesigned as a debriefing activity. To further enhance the potential for in-depth understanding of the behavior of bridges as a construction system, a comprehensive bridge engineering activity could be combined with the shorter duration Bridge Builder™ and Behavior of Spanning Structures activities described below.

Any research-and-experimentation/development activity should engage students in designing, producing, testing, and analyzing. One example is Testing Strength of Concrete Beams (Henak, 1986). Students design and conduct a series of strength tests on model concrete beams. In each test, one variable is changed while other factors are kept constant. Figure 6–9 lists four sample experiments recommended for testing. Usually, students also design and produce valid and reliable testing apparatus. Research-and-experimentation activities often provide numerous opportunities to integrate mathematics, science, economics, composition, and other subjects. Also, they are excellent for allowing students to discover fundamental concepts about the behavior of construction systems.

Three-Phase Activities. Every activity in a course need not include every phase of the model. Some activities may emphasize one or more of the phases to facilitate specific problem-solving skills or the attainment of specific concepts. Three-phase activities can be long or short duration depending upon the concept(s) to be developed, the opportunities for students to repeat phases, and available instruction time. One option is to keep three-phase activities short for quick attainment of concepts followed by a long-duration comprehensive activity where students apply the concepts to the solution of more complex problems. Three-phase construction activities are described below.

Behavior of Spanning Structures (Appalachian Technology Education Consortium, 1992) provides examples of how short three-phase activities focusing on concept attainment can be followed by long-duration comprehensive activities focusing on in-depth understanding. Behavior of Spanning Structures is a module with a series of short activities designed to facilitate the development of technological literacy through the integration of mathematics and science. The problems students solve promote concept attainment leading to a predictive understanding of the behavior of beams

VARIABLES (Changes)	CONSTANTS (Stay the same)
1. Curing time (7, 14, and 21 days)	1. Mixture, reinforcement, curing conditions, beam size, etc.
2. Placement of rebar (top, middle, or bottom of the beam)	2. Mixture, curing conditions, beam size, etc.
3. Texture on the surface of reinforcement steel (Smooth bar-clean, rebar-clean, smooth bar-painted, smooth bar-rusted, rebar-rusted)	3. Mixture, rod size, curing time, curing conditions, beam size, etc.
4. Curing conditions (keep beam damp, let beam dry naturally, dry beam quickly without heat)	4. Mixture, reinforcement, curing time, beam size, etc.

Figure 6–9: Recommended variables and constants for testing concrete beams (Henak, 1986).

and trusses. According to the module developers, "technological literacy requires reasonable competence at the higher predictive conceptual level, namely, those concepts that have to do with 'why' a device, component, or system works or behaves the way it does" (Appalachian Technology Education Consortium, 1992a, p. vi). Several related technology, mathematics, and science concepts are addressed in the module (Figure 6–10).

The five recommended lessons in the module are Beam Loading, Tension and Compression, Stability of Geometric Shapes, Experimenting with Forces, and Depth-to-Span Ratio. Each lesson includes hands-on problem-solving activities. The Experimenting with Forces lesson includes one example of a three-phase activity that includes producing, testing, and analyzing. Students produce a truss from a kit, test each truss member for tension and compression forces, and analyze the behavior of the truss under load conditions. Follow-up activities challenge students to apply their understanding of tension and compression forces, the behavior of trusses under load, and other concepts presented in the module to the design of

Technology Concepts	Mathematics Concepts	Science Concepts
Beam	Best-Fit Line	Force
Bending Moment	Extrapolation	Isotropic Materials
Center of Gravity	Linear Relations	Modulus of Elasticity
Compression	Linear Equation	
Depth-of-Span Ratio	Percentages	
Durability	Perpendicular	
Dynamic Load		
Load		
Material Strength		
Shear		
Stability		
Static Load		
Structural Efficiency		
Structural Integrity		
Tension		
Torsion		

Figure 6–10: Concepts addressed by the behavior of spanning structures module (Appalachian Technology Education Consortium, 1992a).

efficient truss structures. One recommended follow-up activity for high school students is an example of a comprehensive activity. Students are challenged to design and construct a six-foot-truss bridge using lumber struts, coated steel-cable ties, and a plywood deck. The bridge is tested for live-load capacity and analyzed mathematically for structural efficiency.

The short-duration three-phase activities in the spanning-structures module provide students with the conceptual understanding of the behavior of beams and trusses needed to solve the six-foot-truss-bridge problem. Solving the bridge problem reinforces and provides greater understanding of the concepts learned. Another module from the Appalachian Technology Education Consortium, Strength-to-Weight Ratios of Geodesic Domes (1992b), is similar in organization. A series of short-duration three-phase activities address concepts related to the behavior of geodesic domes. Follow-up activities focus on application problems designed to reinforce understanding.

Bridge Builder (Rutherford, 1990) is a low-cost simulation-software program that can be used for a short- or long-duration three-phase activity. Bridge Builder challenges students to design a bridge using a computer, test the strength of the bridge on the computer, and analyze the results. The main objectives of the activity are to introduce students to concepts related to structural engineering, tension and compression, computer-aided drawing and design, and computer modeling and simulation. Bridge Builder engages students in three phases — designing, testing, and analyzing. They do not produce a three-dimensional prototype bridge for testing, which the software simulates. This results in a short-duration activity. However, if students have several opportunities to redesign their bridge after analyzing test results, a long-duration activity will result. Whether a teacher uses the three-phase format for a long- or short-duration activity depends upon the main concepts of the activity. Again, this activity could be followed by the six-foot bridge problem or the bridge-engineering activity to promote greater understanding.

Rover's Rad Pad is a project recommended in the Introduction to Construction Technology (IITEC, 1992d) curriculum guide. The complete plans for the structure, which is pre-designed, are provided in the guide. Following the recommended course outline, students produce (construct) the structure, which limits the problem-solving potential of the activity. If the students are given the plans, a three-phase activity could still be conducted by adding testing and analyzing. Before producing a prototype structure, students could make mock-ups and analyze tests designed to answer these types of questions:

- What size dog will fit through the door opening?

- Will the largest dog that fits through the door be able to enter completely and turn around?

- Is the front stoop high enough to prevent rain or snow from entering and wetting bedding material?

- Is the pitch of roof adequate to provide drainage?

- Is the design cost-effective? Could the homes be constructed and sold for a profit?

- What do potential customers think about the aesthetics of the design?

Answering these questions will lengthen the activity and reduce the amount of content that can be covered. The entire focus of the activity will be changed from a production skills-developing industry approach to a problem-solving-based research-and-experimentation approach. Also, if the students decided to redesign Rover's Rad Pad after analyzing test results, a comprehensive activity will result.

Two-phase Activities. Two-phase activities are short-duration activities, engage students in two of the four phases, and focus on the attainment of individual concepts. Quick attainment of individual concepts is the key feature of a two-phase activity. Three examples of two-phase activities are explained below.

Beam Loading is a two-phase activity in the Behavior of Spanning Structures Module (Appalachian Technology Education Consortium, 1992a) described earlier. In this activity students test and analyze the load-bearing capacity of a simple beam (a yardstick). Design and production are not part of the activity. Loads of one, two, and four pounds are applied to the center of the yardstick beam with a spring scale (Figure 6–11). The amount of sag in the beam is measured and graphed. Then, students predict the behavior of the loaded beam by extrapolating from the data on the graph for a six-pound load. The actual sag is measured and compared to the prediction. The beam-loading activity can be completed in one class period. It focuses on helping students use mathematics (graphing and extrapolation) to understand and predict the behavior of beams under load. As done in the Behavior of Spanning Structures Module, short-duration three-phase activities and long-duration comprehensive activities follow this short-duration activity to provide students with greater understanding through application.

Bridge designers must understand span, sag, and thrust relationships in a cable. Cables are non-rigid structural elements most commonly associated with suspension bridges. As the span of a cable increases and the sag decreases, thrust increases. Students can be given the problem to demonstrate this relationship. In one class period, tests can be conducted and

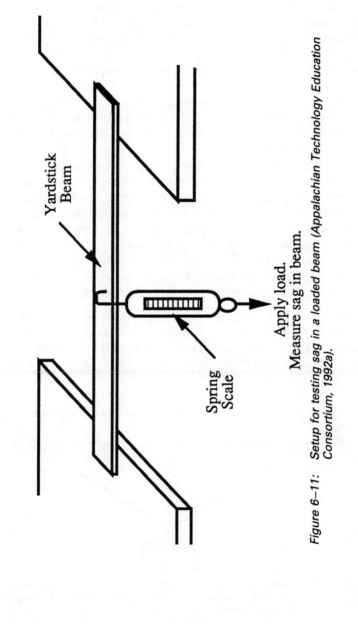

Yardstick Beam

Spring Scale

Apply load.
Measure sag in beam.

Figure 6–11:   *Setup for testing sag in a loaded beam (Appalachian Technology Education Consortium, 1992a).*

analyzed with a cable (rope or string), a load (a weight), and two spring scales (Figure 6–12). Applying a load to the center of the cable suspended with a short span between the spring scales produces a small thrust reading on the scales. Increasing the span and decreasing the sag causes the thrust reading to increase. Mathematics can be integrated by demonstrating that "as the sag is reduced by half, the thrust will increase by a factor of two" (Horton, et al., p. 118). The optimum cable sag can be shown to be equal to one-half the span, which will create a 45-degree angle between the sagging cable and horizontal.

As with other short-duration two- and three-phase activities, this activity focuses on quick attainment of basic structural concepts. Also, a longer-duration comprehensive activity could follow to reinforce conceptual understanding through application.

Low-cost architectural computer-aided drawing and design software programs can be used for two-phase problem-solving activities that focus on designing and analyzing. Students can be given a home or office design problem and specifications from a client for general style, space needs, furniture needs, price range, and special features. Designs can be created on a computer and analyzed by the client. Depending upon the objectives of the activity, the degree of understanding desired, and the time available in the class, the activity could end after analyzing the first design or the space could be redesigned and reanalyzed.

One-phase Activities. Activities that use only one phase of the technological problem-solving method usually focus on analyzing. Designing, producing, or testing in isolation of one or more of the other phases is more related to industrial arts or vocational education than technology education.

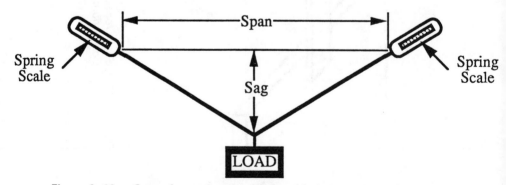

Figure 6–12:   *Setup for testing the relationship between sag, span, and thrust in a spanning cable (Horton, Komacek, Thompson, & Wright, 1991).*

For example, designing a structure to develop architectural drafting skills and not analyzing the design is a common industrial arts project. Similarly, constructing a storage shed designed by the teacher and not testing or analyzing its costs, strength, durability, consumer appeal, or other factors is vocational training. One-phase construction activities should focus on analysis. In a community planning course the environmental, societal, and political impacts of construction could be analyzed. In an architecture-and-construction course, clients' spatial requirements, the need for handicapped access, and the application of design principles in structures could be analyzed. In any construction course, economic aspects could be analyzed. Analyzing activities involve students in researching, reviewing, discussing, diagnosing, calculating, and interpreting problems. In a hands-on technology education course, most one-phase-analyzing activities should be of short duration. Three examples of one-phase activities are explained below.

As explained in module three of Indiana's Community Planning course "before any planning/design project can be successful, critical information that impacts decisions and the decision-making process must be gathered, edited, clearly presented, and objectively analyzed" (IITEC, 1992a, p. 3–1). The main activity recommended in the module to gather critical information is a Resource Opportunity Survey. Student-teams conduct surveys of problems and potentials in the local community. Some survey items relate to trends and conditions of population and business growth, traffic, parking, pedestrian facilities, utilities, condition of buildings, land usage, and availability of services. Survey teams analyze survey data and present findings to the class. The class analyzes the presentations and identifies a list of key concerns for further investigation. The list of key concerns could decide the focus of the remainder of the course.

Engineers calculate loads because the foundation of a structure must support dead and live loads without excessive or uneven settling. Skyscrapers can exert tremendous dead loads on a foundation. Students can analyze the dead loads of a skyscraper by calculating approximations of area, volume, and weight. Here is a sample problem: "A skyscraper 100 stories high has 12 inch thick reinforced concrete floors measuring 30 feet square. The design calls for 36 foundation columns. What is the dead load exerted on each foundation column?" (Horton, et al., 1991, p. 135). The problem is greatly simplified. Dead-load weights for walls, the roof, and other structural components are not included. The solution to the problem is found by analyzing the weights and densities of construction materials. Mathematics is integrated to calculate the dead load applied to each column. In a structural-engineering course, students could perform more detailed analyses on actual structures in the local community. As a follow-up comprehensive activity, model structures could be designed and constructed. The effect of dead loads from the structure on foundation settling could be tested and analyzed.

Analyzing construction costs is essential in all phases of construction. Rough estimates can be calculated for single-family homes using costs per square foot of living space. More detailed estimates of material costs can be made by reviewing working drawings and specifications. Estimating labor costs is more difficult and requires an understanding of the number of different types of workers required and the time required to do each job. Analyzing material and labor costs addresses the economic impacts of construction.

SUMMARY

This chapter addressed three basic questions about high school technology education: (1) why study construction?; (2) what should students study about construction?; and (3) how should construction be taught?

Why study construction? First, construction is a basic technology and an integral part of our technological society. Advanced construction systems require citizens who can understand the behavior of those systems and make informed decisions about their implementation. Second, construction is economically significant. Construction industries employ millions of workers. Hundreds of billions of dollars are spent annually on construction and related economic activity. Finally, as an academic area of study, construction provides excellent opportunities to integrate mathematics, science, and other subjects.

An answer to the question "What should high school students study about construction?" should begin with an analysis of the students' needs. High school is a stressful period. Much of the stress is related to dealing with variations in development and preparing for life after high school. These variations necessitate diverse technology course offerings. Four construction technology courses were proposed to address the varying needs of high school students. Construction Systems would meet the needs of at-risk students and students interested in developing vocational skills without attending vocational schools. Architecture and Construction would meet the needs of students interested in design, drawing, and the visual and spatial characteristics of structures. Structural Engineering would meet the needs of students interested in the technology, mathematics, science, and engineering principles that explain why buildings stand up. Community Planning would meet the needs of students interested in the interactive behavior of construction, social, economic, political, and environmental systems.

The most important question for high school teachers may be "How should construction be taught?" The technological problem-solving method was presented as a model for organizing student learning activities in construction. Comprehensive activities would engage students in designing, producing (constructing), testing, and analyzing structures. Other activities would involve students in one, two, or three phases of the model; depending upon the time available and concepts to be covered. Several construction technology learning activities were presented using the technological problem-solving method.

REFERENCES

American Association for the Advancement of Science. (1989). *Project 2061: Science for all Americans.* Washington, DC: American Association for the Advancement of Science.

Appalachian Technology Education Consortium. (1992a). *Behavior of spanning structures module.* (Available from Kelvin Electronics).

Appalachian Technology Education Consortium. (1992b). *Strength-to-weight ratios of geodesic domes module.* (Available from Kelvin Electronics).

Blankenbaker, E.K. (1982). Construction technology. In T. Wright (Ed.), *Symposium III Proceedings* (pp. 31–34). Muncie, IN: Ball State University.

Botvin, G.J., Baker, E., Dusenbury, L., Tortu, S., & Botvin, E.M. (1990). Preventing adolescent drug abuse through a multi-modal cognitive-behavioral approach: Results of a 3-year study. *Journal of Consulting and Clinical Psychology. 58*(4), 437–446.

Boyer, E.L. (1983). *High School: A report on secondary education in America.* New York: Harper & Row.

Business Council for Effective Literacy. (1991). Construction and basic skills. *BCEL Newsletter for the Business and Literacy Communities.* April 1, 9–11.

Construction Industry Workforce Foundation. (1990). *Perceptions and attitudes of young people about the construction industry.* Author.

Dugger, W.E., Jr., French, B.J., Peckham, S., & Starkweather, K.N. (1992). Seventh annual survey of the profession: 1990–91. *The Technology Teacher. 51*(4), 13–16.

Edmison, G.A. (1992a). *Approaches: Teaching strategies for technology education.* Reston, VA: International Technology Education Association.

Edmison, G.A. (1992b). *Delivery systems: Teaching strategies for technology education.* Reston, VA: International Technology Education Association.

Farrell, E. (1990). *Hanging in and dropping out: Voices of at-risk high school students.* New York: Teachers College Press.

Gasperow, R.M. (1988). The future need for skilled workers. *Constructor, 70*(10), 12–14.

Gaylord, E.H., Jr., & Gaylord, C.N. (Eds.). (1990). *Structural engineering handbook.* New York: McGraw-Hill.

Harper, J.F., & Marshall, E. (1991). Adolescents' problems and their relationship to self-esteem. *Adolescence. 26*(104), 799–808.

Hatch, L. (1988). Problem solving approach. In W. Kemp & A. Schwaller (Eds.), *Instructional strategies for technology education* (pp. 87–98). Mission Hills, CA: Glencoe.

Havighurst, R.J. (1972). *Developmental tasks and education.* New York: David McKay.

Hayes, I.D. (1992). *Welcome address: National Institute for Material Science and Technology.* Battelle-Pacific Northwest Laboratories.

Henak, R. (1986). Testing strength of concrete beams. # CO-P-005, Center for Implementing Technology Education. Muncie, IN: Ball State University.

Henak, R. (1991). Construction courses for the 21st century. *School Shop/Tech Directions. 50*(7), 9–11.

Henry, W.R. (1989). AGC President Paul Emerick. *Constructor, 71*(4), 16–19.

Hoborman, H.M., & Garfinkel, B.D. (1988). Completed suicide in children and adolescents. *Journal of American Children and Adolescence. 27*(6), 689–695.

Home Builders Institute. (1991). *An analysis of America's construction industry workforce & occupational projections: 1990–1996.* Washington, DC: Home Builders Institute.

Horton, A., Komacek, S.A., Thompson, B.W., & Wright, P.H. (1991). *Exploring construction systems: Designing, engineering, building.* Albany, NY: Delmar.

Indiana Industrial Technology Education Curriculum. (1992a). *Community planning.* Muncie, IN: Center for Implementing Technology Education, Ball State University.

Indiana Industrial Technology Education Curriculum. (1992b). *Constructing structures.* Muncie, IN: Center for Implementing Technology Education, Ball State University.

Indiana Industrial Technology Education Curriculum. (1992c). *Construction planning and design.* Muncie, IN: Center for Implementing Technology Education, Ball State University.

Indiana Industrial Technology Education Curriculum. (1992d). *Introduction to construction technology.* Muncie, IN: Center for Implementing Technology Education, Ball State University.

Johnson, C.A., Pentz, M.A., Weber, M.D., Dwyer, J.H., Baer, N., MacKinnon, D.P., Hansen, W.B., & Flay, B.R. (1990). Relative effectiveness of comprehensive community programming for drug abuse prevention with high-risk and low-risk adolescents. *Journal of Consulting and Clinical Psychology. 58*(4), 447–456.

Johnson, J.R. (1992). Technology education: An imperative. *The Technology Teacher. 52*(2), 3–5.

Kemp, W.H., & Schwaller, A.E. (Eds.). (1988). *Instructional strategies for technology education.* Mission Hills, CA: Glencoe.

Kitzmann, R. (1988). *Classroom activities in construction: Technology education.* Madison, WI: Wisconsin Department of Public Instruction.

Kleinfeld, L.A., & Young, R.L. (1989). Risk of pregnancy and dropping out of school among special education adolescents. *Journal of School Health, 59*(8), 359–361.

Komacek, S.A., & Bolyard, G. (1992, in press). The technological problem-solving method applied to transportation. In L. Litowitz & J. McCade (Eds.), *Proceedings of Technology Symposium XIV: Teaching Content versus Problem Solving.* Millersville, PA: Millersville University of Pennsylvania.

Lester, D. (1991). Social correlates of youth suicide rates in the United States. *Adolescence. 26*(101), 55–58.

Lindgren, H.C., & Suter, W.N. (1985). *Educational psychology in the classroom.* Monterey, CA: Brooks/Cole.

Maley, D. (1985). Keynote address: Issues and trends in technology education. In N. Andre & J. Lucy (Eds.), *Proceedings of technology education Symposium VII: Technology education: Issues and trends* (pp. 3–14). California, PA: California University of PA.

Merritt, F.S. (1982a). Building systems. In F.S. Merritt (Ed), *Building design and construction handbook* (pp. 1–1 to 1–37). New York: McGraw-Hill.

Merritt, F.S. (1982b). Structural theory. In F.S. Merritt (Ed), *Building design and construction handbook* (pp. 5-1 to 5-173). New York: McGraw-Hill.

Mullin, M.H. (1991). *Educating for the 21st Century: The challenge for parents and teachers.* Lanham, MD: Madison Books.

National Center for Health Statistics. (1989). *Vital statistics of the United States, 1987: Volume 2 Mortality, Part B.* Washington, DC: U.S. Government Printing Office.

New York State Education Department. (1990) *Technology education: Introduction to technology; Grades 7 & 8.* Albany, NY: New York State Education Department, Division of Occupational Education Programs.

Newmann, F.M., & Behar, S.L. (1982). *The study and improvement of American high schools: A portrait of work in progress.* Madison, WI: Wisconsin Center for Education Research.

Page, R., Clarke, R., & Poole, J. (1982). *Modular courses in technology: Problem solving.* Edinburgh, Scotland: Oliver & Boyd.

Rhodes, J.E., & Jason, L.A. (1990). A social stress model of substance abuse. *Journal of Consulting and Clinical Psychology. 58*(4), 395-401.

Rutherford, B. (1990). *Bridge Builder* [Computer program]. Aberdeen, WA: Shopware Educational Systems.

Salvadori, M., & Tempel, M. (1983). *Architecture and engineering: An illustrated teacher's manual on why buildings stand up.* New York: The New York Academy of Sciences.

Savage, E., & Sterry, L. (1990). A conceptual framework for technology education. *The Technology Teacher. 50*(1), 6-11.

Seymour, R.D. (1987). A model of the technical research project. In E.N. Isreal & R.T. Wright (Eds.), *Conducting technical research* (pp. 46-57). Mission Hills, CA: Glencoe.

Shafer, D., Garland, A., Gould, M., Fisher, P., & Trautman, P. (1988). Preventing teenage suicide: A critical review. *Journal of the American Academy of Child & Adolescent Psychiatry. 27*(6), 675-687.

Snyder, J.F., & Hales, J.A. (Eds.). (1981). *Jackson's Mill industrial arts curriculum theory.* Charleston, WV: West Virginia Department of Education.

Tang, S.L., & Chin, I.R. (1982). Buildings: General design considerations. In E.H. Gaylord, Jr., & C.N. Gaylord (Eds.), *Structural engineering handbook* (pp. 21–1 to 21–51). New York: McGraw-Hill.

Technology Education Advisory Council. (1988). *Technology: A national imperative.* Reston, VA: International Technology Education Association.

U.S. Department of Labor. (1991). *What work requires of schools: A SCANS report for America 2000.* Washington, DC: U.S. Government Printing Office.

U.S. Department of Labor. (1992). *SCANS in the schools.* Washington, DC: U.S. Government Printing Office.

Waetjen, W. (1989). *Technological problem solving.* Reston, VA: International Technology Education Association.

West Virginia Department of Education. (1987). *Construction curriculum guide.* Charleston, WV: WV Department of Education.

Wood. (1987). *Exploring construction technology.* Stillwater, OK: Mid-America Vocational Curriculum Consortium.

Wynn, R.H. (1987). A sophomore-level structural design experience. *Engineering Education.* 77(4), 231–233.

Zelnik, M., & Shah, F. (1983). First intercourse among young Americans. *Family Planning Perspectives.* 15, 64–69.

CHAPTER 7

Preparing Teachers For Construction In Technology Education

Jack W. Wescott, Ph.D.
Department of Industry and Technology
Ball State University, Muncie, IN

There is likely to be more change in teacher education between now and the year 2000 than there has been in the entire history of teacher education. The reason is clear: The demand for more competent teachers in the public schools is being expanded into demands that universities also improve their teacher education programs. The call for excellence in the public schools coupled with the expectation that secondary schools should prepare individuals to work and compete in a highly competitive international arena will continue to increase pressures on teachers.

This chapter is concerned with the pre-service education of technology education teachers. It brings together different dimensions of existing theory and practice and provides input into how teacher education programs can be improved. Where appropriate, specific inferences directed towards the area of construction in technology education will be included to address the question of how best to prepare teachers of construction in technology education.

THE VISION

In his book entitled, *Schools of Quality: An Introduction to Total Quality Management in Education*, Bonstingl (1992) describes a vision as follows:

> Where do we want to be in the future? The ideal condition for the profession, even if it seems out of reach. It describes the preferred

future and what we see in that future. Finally, the vision assumes that all resources are available to achieve the vision. (p. 52)

Therefore, it seems appropriate that a discussion of a teacher education program for construction in technology education should begin with a vision.

This vision should take into consideration two guideposts. First, the program should be guided by a vision of the attributes of an outstanding teacher of construction in technology education. Such a vision establishes the philosophical foundation for program development. The second guidepost is a consistent awareness of the persistent criticisms of traditional teacher preparation programs with implications for the future. A review of contemporary reports on the condition of higher education can provide an intuitive understanding of the major criticisms of traditional programs and provide insight into new techniques for conceptualizing and implementing effective teacher education programs.

The ideal teacher of construction in technology education is a professional who is foremost an expert in classroom and laboratory instruction. The primary vision of a pre-service teacher education program should be to prepare outstanding teachers. In order to keep this vision in focus, it is important to make basic assumptions about the nature of the ideal teacher of construction in technology education.

CHARACTERISTICS OF AN EFFECTIVE TEACHER OF CONSTRUCTION IN TECHNOLOGY EDUCATION

Andrew (1989) organized the basic assumptions of effective teachers into the necessary and contributory conditions for good teaching. The four necessary conditions for an effective teacher are considered to be the basic requirements for most good teachers. These conditions are often established in the individual prior to beginning a teacher education program. He stated that the ideal beginning teacher should:

1. Possess above-average academic skills.

2. Possess a strong general education.

3. Possess sufficient depth in a special subject field to give confidence, credibility, and the background for further graduate work or other career options if desired.

4. Provide evidence of teaching potential, commitment, and appropriate interpersonal skills for successful teaching.

"These skills have been defined to include the ability to communicate effectively with children and adults, good listening skills, sensitivity to the needs of others, and the ability to work positively with children and adults" (Andrew, 1989, p. 45).

Andrews (1989) also identified contributory conditions which are those factors that will enhance teaching. They are the areas of cognitive, pyschomotor, and affective learning that are the major concerns of the professional program in teacher education. In this regard he stated that an ideal beginning teacher should:

1. Work from an understanding of children that emphasizes a knowledge of human development and learning theories.

2. Have available a variety of proven teaching models and be assisted in the development of an educational philosophy and effective personal style of teaching. This condition is supported by our knowledge of the wide range of effective teaching methods, the documented value of flexibility in teaching style, and our vision of a teacher as one who has a clear sense of professional identity as well as effective teaching practices.

3. Possess a knowledge of the structure of public education and procedures for effecting change in that structure.

4. Possess a knowledge of the significant assumptions and philosophical points of view that underlie teaching and schooling and should have defined a personal philosophy of education (p. 46).

It is also important to review the research relative to the characteristics of a successful technology education teacher. In a survey study entitled Characteristics for Successful Professional Performance as a Technology Teacher, Lacroix (1987) asked 144 technology education teachers three questions: (1) What are the characteristics (excluding technical skills) required for successful professional performance as an industrial arts/ technology education teacher?; (2) Can these characteristics be taught or developed during the teacher education program?; and (3) Should there be criteria for the selection of teacher education students? The results of the study were summarized as follows:

1. There is probably very little difference between effective teachers of technology education and effective teachers in any other area of study.

2. Teacher personality probably plays some role in the overall make-up of the effective teacher.

3. The effective teacher probably holds the following variables in high regard and uses them more often than does a less effective teacher: (a) clarity, (b) variability, (c) enthusiasm, (d) task orientation, (e) provision of student opportunity to learn critical materials, (f) patience and truly cares about students, and (g) organization and preparation (p. 35).

Similarly, Lacroix indicated that these characteristics should be addressed in teacher education programs. He states:

Teacher education programs address more than the development of "technical skills." Teacher education programs also deal with the teaching of those "nontechnical skills" that help to make an effective teacher—a person who deals with people. (p. 38)

More recently, a report published by the U.S. Department of Labor (1992) entitled, *Learning a Living: A Blueprint for High Performance* attempted to define the competencies necessary for a high performance workplace. A survey of employers, supervisors and especially employees across the spectrum of the American economy found a clear pattern of requirements. The survey concluded that there were five know-how competencies and a three part foundation of skills and personal qualities that are needed for solid job performance. The competencies and skills listed below were stated in the SCANS Report (p. 6).

Workplace Competencies:

- **Resources** - They know how to allocate time, money, materials, space, and staff.

- **Interpersonal skills** - They can work on teams, teach others, serve customers, lead, negotiate and work well with people from culturally diverse backgrounds.

- **Information** - They can acquire and evaluate data, organize and maintain files, interpret and communicate, and use computers to process information.

- **Systems** - They understand social, organizational, and technical systems; they can monitor and correct performance; and they can design or improve systems.

- **Technology** - They can select equipment and tools, apply technology to specific tasks, and maintain and troubleshoot equipment.

Foundation Skills:

- **Basic skills** - reading, writing, arithmetic and mathematics, speaking and listening.

- **Thinking skills** - the ability to learn, to reason, to think creatively, to make decisions, and to solve problems.

- **Personal qualities** - individual responsibility, self-esteem and self-management, sociability, and integrity.

It is interesting to note that these workplace competencies and foundation skills have direct implications for a teacher of construction in technology education. Also, it is difficult to argue that the characteristics identified by the report are not the desirable characteristics of a construction in technology education teacher. Therefore, it may be appropriate to look outside the traditional walls of pre-service teacher education for other valuable input regarding the characteristics of an effective teacher of construction in technology education.

NATIONAL CALLS FOR REFORM

Traditionally, teacher education has been criticized whenever important interest groups decide that something is wrong with education in general. In the last several years the education community has been inundated by a series of education reform reports, beginning with *A Nation at Risk* (1983) and the *National Commission on Excellence in Education* (1985), to the more recent reports of the Holmes Group, including *Tomorrow's Teachers* a report by education deans (Jacobson, 1986). In addition, a majority of state legislatures, boards, and departments of education have joined the so-called "reform movement" by attempting to increase the credentialing and testing of teachers. As a result, there have been significant calls for noticeable improvement in education, which have directly influenced those institutions that prepare teachers.

An attempt to summarize the recommendation of the reports was conducted by Clark (1986). Three distinct themes seem to be apparent in these reports. The first is a need for a more thorough subject-matter preparation of teachers. Specifically stated, teacher education programs should:

- Limit the number of education courses; expand the role of the liberal arts colleges;

- Provide alternate paths to certification; increase the number of content courses;

- Expand field involvement in program design; and

- Move to a five-year program with the fifth year concentrating on professional education and field experience.

The second major criticism relates to the standards for teacher education candidates. The following solutions have been presented:

- Increase standards for admitting students to teacher education programs.

- Increase performance standards in teacher education programs.

- Increase standards for completion of teacher education programs.

- Add post-program examinations of competency for teacher certification.

A third criticism concerns the financial and social support for teacher training. In response to this problem, the reports have suggested the following:

- Provide funds for targeted recruitment efforts (especially science and mathematics).

- Increase scholarship and loan provisions for teacher education candidates.

TEACHER SELECTION

It seems logical that the process of selecting potential teachers should begin with what is to be considered the major attributes of an effective teacher of construction in technology education. According to Howey and Strom (1987), the professional preparation of teachers should attract individuals who are able to think conceptually and who possess the qualities of being adaptable, inquisitive, critically inventive, creative, self-renewing, and oriented to moral principles. Furthermore, Howey and Strom identify seven generic characteristics of effective teachers, which are appropriate for the teacher of construction in technology education. These are as follows:

1. Form and use multiple concepts; consider different and conflicting perspectives in thinking and decision making.

2. Organize and use ideas in a creative manner in problem solving.

3. Understand and use others' perspectives in communication.

4. Seek belief-practice congruity.

5. Perceive and use conceptions regarding the needs of diverse groups.

6. Evaluate the impact of their actions on other people.

7. Develop a personally meaningful work identity (p. 8).

It also seems appropriate that various measures of academic aptitude and achievement should be employed to ensure that intellectually able persons teach in technology education programs. Wescott (1991) conducted research to determine the relationship between Scholastic Aptitude Test (SAT) scores and the academic success of technology education majors. The findings of the study indicate that some significant relationships do exist between an individual's SAT scores and academic performance. The most significant relationship existed between the combined math and verbal scores and the cumulative grade-point average at the time of graduation. Also, the math score of the SAT provided the most significant relationship relative to academic performance within the major. Although these quantitative measurements were never intended to be the sole criteria for the selection of teacher education candidates, they can provide the teacher educators with valuable information regarding an individual's intellectual ability.

The selection of students for admission into a technology teacher education program is one of the most critical elements in the total process. According to Devier (1981, 1986), the technology education major usually enters a teacher education program due to an affiliation with a middle and/or high school program and shows a very strong interest in industrial arts/technology-type activities such as hands-on, personal interests, and hobbies. Because of this background, many teacher education majors have made significant value-judgements relative to the selection of content and methodology appropriate for technology education. This "teach as I was taught" mentality is not always good, especially if the prospective technology education major completed a traditional, materials-oriented (woods, metals, drafting, power mechanics) industrial arts program. Interestingly, many individuals select a career as a technology teacher based upon an interest in things, such as tools, materials, and process, not because they are interested in working with young people. Therefore, individuals responsible for the recruitment of prospective technology education teachers need to explore those populations that display the characteristics of an effective technology education teacher described previously. DeVore (1975) summarized the type of individual the profession needs to recruit when he stated:

Generally, significant changes in institutions, in disciplines, and various fields of endeavor have usually been made by people outside the institutions or disciplines or by the young or by those new to the discipline. Thus, if we desire to make an impact on education and the future society we must solicit the assistance of those with a new perspective, those outside the discipline, the young and those new to the discipline. That is the story, the rest is commentary. (p. 33)

MAJOR COMPONENTS OF CONSTRUCTION IN A TECHNOLOGY TEACHER EDUCATION PROGRAM

Traditionally, teacher education programs have been organized around three major elements: general education, content specialization, and professional education. Figure 7–1 provides a graphic model of these major elements.

General Education

A good liberal education is an education which liberates the mind from the shackles of prejudice and superstition and the confines of a single culture, that permits one to move freely and joyfully in the past and present and to speculate objectively, with fellowmen about the future—is a foremost aim of our schools. How far short we fall of ideal is probably directly related to how far teachers in the classroom of our schools have themselves fallen short in their liberal education. (p. 4)

This quotation from Robert Bush (1966) is as true in the 1990's as it was in the 1960's. Today's teacher of construction in technology education needs a strong general education background. For years there has been a significant effort to provide an operative definition of the term "liberal arts education" and to define its relationship to general education. As a point of reference, liberal education is defined as those educational experiences that provide the individual a knowledge of the concepts, skills, and values necessary for educated men and women to live purposefully in a modern society. Whatever the definition, candidates for the teaching profession generally acquire their familiarity with the liberal arts through the general education component of their college programs. Although the terms liberal arts education and general education are usually not synonymous, they are somewhat interchangeable when used in reference to teacher education programs.

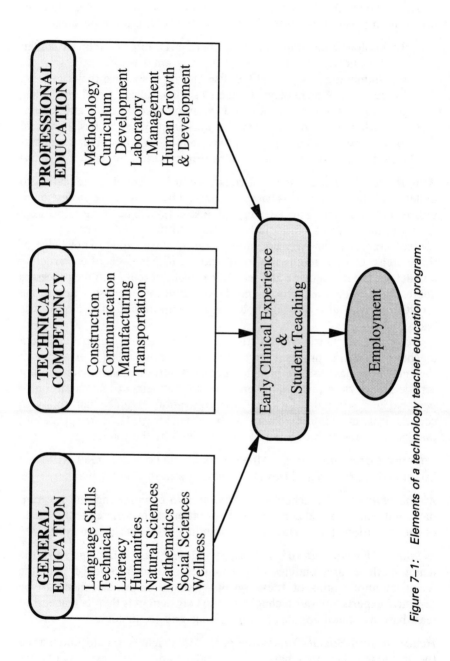

Figure 7–1:   Elements of a technology teacher education program.

The importance of a liberal arts component in a technology teacher preparation program is emphasized by Custer (1990), when he stated:

> Technology education needs the liberal arts. Interdisciplinary configurations need to be developed. Conversation with the social sciences and humanities are critical because of the perspective they can bring to the study of technology. Concepts such as design, problem-solving, experimentation, innovation and invention take on color and dimension when examined against the backdrop of aesthetics, historical experience, ethics and philosophical inquiry. A convergence is necessary if technology is to be examined in its widest dimensions. (p. 54)

Also, the need for the teacher of construction in technology education to understand the liberal arts is based upon a call by many in the profession to utilize an interdisciplinary strategy of instruction. The integration of a variety of disciplines such as math, science, and the social sciences into the content area of technology education requires an adequate knowledge of those areas. Understanding the social and cultural impacts of constructing a structure, the mathematic and scientific principles related to the processes and techniques of construction, and the contribution of construction to the growth of civilization are examples of the importance of a liberal arts background.

According to Petrie (1987), the general education component of a teacher preparation program serves four critical functions: (1) extension of the knowledge base formed in high school, (2) introduction to scientific and artistic modes of inquiry and expression, (3) refinement and extension of personal and societal values, and (4) cultivation of individual ability to communicate in an informed and reflective manner (p. 1). More specifically, the general education of a teacher should include the following:

Effective Communication.   All teachers should be able to read, write and listen; and express themselves in a coherent and intelligible manner.

Mathematics.   All teachers should be able to comprehend and use fundamental mathematical concepts and operations appropriate for their field of specialization. This includes understanding research data analysis.

Scientific Understanding.   All teachers should have a basic grasp of the major methods and findings in the natural and social sciences and the technical implications of these results. Not all individuals need to be technical experts, but our technological society demands that its teachers at least have a general appreciation of science and technology.

Historical and Social Consciousness.   All teachers should understand the includable fact that both our individual and social experiences are historically grounded. To prepare students for a global perspective in the 21st century, teachers need a comprehensive background in the historical and cultural traditions shaping the societies of the world.

Humanities. All teachers should appreciate the human condition as it is illuminated by language, literature, and philosophy, so that they may encourage their students to live meaningful lives (p. 2).

The National Council on Accreditation of Teacher Education (NCATE) *Folio Preparation Handbook* (1989) for technology education makes specific reference to general education courses that are required of all students and are taught by faculty fully qualified in each subject area. The required content areas in general education include mathematics; and life, physical, and social sciences. The NCATE document also specifies the related areas of philosophy, psychology, sociology, economics, anthropology, geography, the arts, music, and the humanities as being important to teacher education majors.

Henak (1991, p. 8) recommended a 48 credit-hour general education structure for the technology education major.

RECOMMENDED 48 HOUR GENERAL EDUCATION STRUCTURE

Language Skills - 9 credits
    Written Communications
    Oral Communications
Technological Literacy - 6 credits
    Man/Society/Technology
    Computer Literacy
Humanities - 3 credits
    Literature
    Fine Arts
Natural sciences - 3 credits
    Applied Physics with laboratory
    Applied Chemistry with laboratory
    Applied Biological Science with laboratory
    Applied Geological Science with laboratory
Mathematics - 6 credits
    Algebra
    Trigonometry
Social Science - 12 credits
    American Government
    Economics

Global Studies/Futuring
World History
Psychology
Sociology
Wellness - 3 credits
Suggested electives:
Applied Physics II with laboratory
Applied Chemistry II with laboratory
Computer Programming
Pre-calculus
Foreign Languages
Technological Literacy

Technical Competence

A contemporary undergraduate program for preparing technology education teachers includes subject matter based upon content organizers. The most widely accepted organization is the construction, manufacturing, communication, and transportation model. The teacher education program is designed to provide the future teacher with knowledge and competencies in each of these areas. Here, the discussion of the technical competence relates directly to the construction cluster.

The content area of construction in technology education can be defined as, *the efficient use of manufactured goods, materials and resources to build a structure on a site*. This definition can be divided into the basic elements of designing, constructing, and using structures. A more detailed breakdown of these basic elements is found in chapter 1. According to the International Technology Education Association and its affiliate Council of Technology Teacher Education (1989) teacher education programs should be comprised of required and well-developed coursework and laboratory experiences in the study of construction and construction systems that:

a. provide, as a minimum, fundamental knowledge of construction and construction systems including: 1) the technological concepts related to the technical means associated with the conceptualization, analysis, design, techniques; 2) procedures used in creating and providing a wide range of constructed edifices related to human shelter, public buildings, streets and highways, bridges, causeways, transportation terminals, recreation facilities, hydroelectric, and other energy conservation systems and facilities for the manufacturing, processing and communication industries; and

b. develop within each student the ability to perform in several technical areas of the construction systems using state-of-the-art instruments, devices, equipment, and materials to design, build, analyze, and evaluate various forms of construction.

In a Council of Technology Teacher Education monograph entitled *Elements and Structure for a Model Undergraduate Technology Teacher Education Program*, (Henak, p. 13, 1991) an outline for the organization of the content related to construction in technology education was provided.

CONSTRUCTION IN TECHNOLOGY EDUCATION

I. Scope
   a. buildings
   b. civil structures
   c. heavy industrial structures
II. Universal systems model
   a. inputs (specifications, resources)
   b. productive process (prepare to build, set foundations, build super-structures, install mechanical systems, enclose superstructure, finish project, complete site, service project
   c. managerial processes
   d. outputs (buildings, heavy industrial structures, civil structures)
III. Managed system
   a. establish the organization
   b. design the structure
   c. prepare to build the structure
   d. build the structure
   e. transfer ownership
   f. control the system
IV. Impacts
   a. individual
   b. societal
   c. environmental
   d. economical

Suggested Technical Course Offerings. The pre-service technical courses in the cluster of construction should be laboratory experiences that focus on technology as a system that integrates human, tools and machines, and material resources to produce structures and services that extend

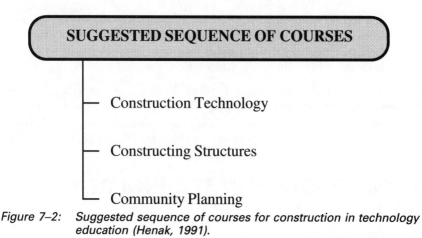

SUGGESTED SEQUENCE OF COURSES

— Construction Technology

— Constructing Structures

— Community Planning

Figure 7–2:   Suggested sequence of courses for construction in technology
education (Henak, 1991).

human potential. Simply stated, the prospective teacher should be able to know, do, and value construction as it relates to the study of technology (Wright, 1990).

The number of credit hours in each of the selected clusters is usually limited due to the fact that technology teachers are required to develop competencies in each of the clusters. It is important to note that the sequence of courses suggested below does not constitute a major. Figure 7–2 represents a suggested sequence of courses for a teacher education program. Each of the suggested courses is further discussed in terms of using course descriptions and learning outcomes.

Course Title: Construction Technology
Course Description: Focuses on construction and management practices used to design, engineer, and construct residential, commercial, industrial, and civil structures. (3 semester hours)
Instructional Objectives:

1. Develop a basic understanding of the processes, procedures, and resources related to the building of a structure on site.

2. Develop an understanding of the social, cultural, and environmental issues related to construction technology.

3. Develop a basic understanding of the design, planning, and managing processes for constructing a structure.

4. Develop knowledge related to the selection and application of basic construction materials.

5. Select, identify, and service the major mechanical systems of a structure.

6. Safely perform basic operations with construction tools, equipment, and materials.

Course Title: Constructing Structures
Course Description: Comprehensive study of the activities involved in preparing to build, construct, and complete residential, commercial, industrial, and civil structures.
Instructional Objectives:

1. Develop an understanding of the basic concepts related to the design and planning of a structure.

2. Safely perform basic operations with construction tools, equipment, and materials related to the process of constructing a structure.

3. Develop skills and knowledge of the major systems of a structure.

4. Select, develop, and organize construction laboratory experiences appropriate for public school education.

5. Understand the social, cultural, and environmental issues related to construction technology.

6. Develop problem-solving skills related to the design and fabrication of structures.

Course Title: Community Planning (3).
Course Description: Introduces students to the activities involved in planning communities. Emphasis is placed on construction, demand, and site factors as related to developing housing, streets, transportation systems, parks, shopping areas, and recreational centers based on public need.
Instructional Objectives:

1. Describe the areas of concern of community planners and the procedure and tools they use to produce community development proposals.

2. Describe community planning in terms of the needs and the rights of people to maximize their quality of life.

3. Prepare a written community development proposal that includes plans for land use, housing, transportation, public facilities, economic development, environmental considerations, and fiscal budgeting.

4. Identify and develop the major components of a community development proposal.

Instructional Faculty. The technical courses in teacher preparation courses should be taught by an experienced teacher education faculty. However, this is generally not the case, as most teacher education institutions have undergone a significant shift in emphasis from predominantly teacher preparation programs to industrial technology programs that prepare individuals to work in industry. The reasons for this shift are numerous and not necessarily pertinent here. Nonetheless, a large percentage of the teacher education faculty have chosen to abandon teacher preparation programs to teach in industrial technology programs. Householder (1992) further describes the problem as follows:

> Few institutions in the United States are currently capable of offering a separate undergraduate technology teacher education program. Few departments are able to exert academic control over an entire technology teacher education major. Few programs are able to recruit and develop a well-prepared faculty with the requisite technical specialties to teach the required courses in the major. Even fewer programs have the financial resources to provide contemporary laboratories in all the relevant technical specialties. (p. 4)

The retrained teacher education faculty has also been supplemented by faculty with industrial and/or engineering backgrounds. This shift in program and faculty, combined with low teacher education enrollments, has forced many departments to use industrial technology courses and faculty to deliver the technical competencies to the teacher education majors. This situation does not always provide the teacher education student an understanding of the content, methodology and philosophy appropriate for a teacher of construction in a public school technology education program.

Professional Education

The professional education element is usually taught by the major department in conjunction with the school of education. The presentation of the professional education element can be organized into the three phases of exploring teaching, professional coursework, and student teaching.

Phase One: Exploring Teaching. Early in the teacher education program (freshman or sophomore year), the prospective technology education teacher should be provided with experiences that develop an understanding of the nature of teaching. The purpose of these experiences is to make certain that the prospective teacher has an accurate and realistic perception of teaching. At times, students have become discouraged about their career choice after they have student-taught and completed all of the coursework required for graduation. Early participation in these courses can provide a realistic orientation for the beginning technology education teacher.

Some of these early exploratory experiences can be effectively delivered in a classroom setting. Suggested topics for such a course include: characteristics of good teachers, teaching technology, terminology related to technology education, and professionalism. However, it becomes necessary at some point in time to allow the students to interact in a public school setting as a teacher. Goodlad, in an interview with Brandt (1991) described the importance of the relationship between teacher preparation and the public schools, when he stated: "Any teacher education program conducted without collaboration of schools is defective. And I mean collaborative where schools are equal partners" (p. 13).

During the early field experience, students should be placed in a public school setting under the direct supervision of a master technology educator. The emphasis of this experience is on participation rather than observation. Students should be encouraged to assume teaching tasks immediately (e.g., giving lessons, monitoring student progress, and tutoring of students). The early field experience is important, as it allows the student to develop a realistic viewpoint of their career choice, and assists in the application of the instructional concepts presented during the technical and professional coursework. Surveys of teacher preparation programs at the national level have revealed an increasing amount of firsthand experiences required of teacher education majors (Heath, 1984).

Phase Two: Professional Education. The second phase of the program normally begins during the first semester of the junior year. This sequence of courses is designed to provide learning experiences related to the major areas of instructional methods, curriculum development, laboratory management, and learner assessment. Also, a review of contemporary educational literature stresses the need for future teachers to develop skills related to problem solving, cooperative learning, and the development of interdisciplinary learning experiences.

A listing of suggested professional courses in professional education was identified by the Undergraduate Studies committee of the Council of Technology Teacher Educators (Henak, 1991, pp. 21-22). This listing of courses includes the following:

Course Title: Teaching Technology (First or second level course)
Course Description: Students work individually and in collaborative groups to investigate a career in teaching technology, begin developing a philosophy of

teaching relevant to technology, and gain an understanding of the historic development, terminology and curriculum development process.
Topics:

1. Describing a career in teaching technology

2. History of Technology

3. Differentiating terminology in the field of technology

4. Developing a philosophy of teaching

5. Describing curriculum development process used in technology

Course Title: Methods for Teaching (300 level course)
Course Description: Problems in developing and implementing instructional strategies, preparing assessment instruments, managing the physical environment, and maintaining a cooperative classroom environment. Topics developed in the context of field experiences whenever possible.
NOTE: This course should be taught in the major department with the credit hours used to satisfy the professional requirements.
Topics:

1. Describing the characteristics of a group of learners

2. Communicating the rationale, objectives, structure, and intended outcomes for a technology course

3. Developing an instructional strategy for teaching problem solving through a problem-solving approach in which technical content and activities are emphasized

4. Designing an assessment system for a technology course

5. Designing and managing the physical environment for a technology course

6. Managing classroom behavior

7. Producing instructional media

8. Teaching value/moral issues within a technology context

9. Teaching futuristics in the technology classroom

10. Teaching students with special needs

11. Developing and implementing a safety program in technology laboratories

Course Title: Curriculum Development in Technology: (400 level course)
Course Description: Emphasizes contemporary approaches to technology, curriculum development, and the preparation and use of instructional media. Students plan and execute a teaching unit in a public school setting.
Topics:

1. Developing criteria for evaluating technology curriculum materials.

2. Evaluating and selecting instructional units in technology.

3. Developing and implementing instructional units in technology.

4. Preparing and using instructional media.

Phase Three: Student Teaching. The student teaching experience is the capstone of the teacher preparation experience. Therefore, it is imperative that students be placed in a quality public school technology education program that is taught by a master teacher. Whenever possible, students should not use the same site for both student teaching and early field experience. Furthermore it is recommended that students not be placed in a school that they attended previously. NCATE requires that this be a carefully monitored and supervised full-time experience. It is also recommended that the program is administered and supervised by a qualified technology education program faculty in cooperation with a master teacher in an approved technology education program.

In summary, the desirable characteristics of a sequence of teacher education courses for construction in technology education is as follows:

1) Provides laboratory experiences that present the students with the opportunity to reinforce abstract concepts with concrete experiences.

2) Applies major principles to other disciplines (general education) to the study of construction.

3) Develops the student's ability to apply construction-related tools, materials, machines, processes, and technical concepts safely and efficiently.

4) Develops problem-solving and decision-making abilities relative to construction materials, resources, and processes.

5) Develops students' attitudes, knowledge, and skills regarding how construction systems function.

PROGRAM ASSESSMENT

In recent years, a new field of study has emerged within the social and behavioral sciences. This new field is assessment, and education has actively implemented its methodology. Nearly every local, regional, or national meeting of educators has a portion of its program devoted to assessment issues. The pressures of accountability, consumerism, and financial restraints are examples of some of the contemporary forces that have led to the demand to evaluate people and programs. Teacher education has not escaped the increased interest in assessment. Pressures to assess teacher education programs have come from many sources, most of which are identified in a previous section entitled National Calls for Reform.

Initially, it is important to define the term program assessment. Stufflebeam (1982) defines teacher education program evaluation as follows:

> Evaluation is the process of delineating, obtaining, and applying descriptive and judgmental information concerning the worth and merit of some program's goals, design, implementation, and impacts in order to promote improvement, serve needs of accountability, and foster understanding. (p. 138)

The evaluation of a teacher education program serves two major purposes. First, it allows individuals to assess the quality of their program in terms of faculty, curriculum, facilities, and graduates. Secondly, and most importantly, is that the evaluation process provides objective input regarding the modification of existing programs.

Approaches To The Evaluation Of Teacher Education

Follow-up Studies. In this approach recent graduates are sent a questionnaire or participate in a structured interview to identify general opinions about the program and to make suggestions for program modification. According to Raths, et al. (1981), the information gathered from such procedures is rarely seen as useful in making changes in programs because the findings of follow-up studies are rarely course-specific. That is, the specific advice of a graduate is often accepted by the faculty as valid, but under the pretense that it be covered in someone else's course.

The Accreditation Process. Another standard approach to evaluation of teacher education programs is the accreditation process. In the field of technology education, NCATE has played a major role in the evaluation of teacher education programs. The importance of the evaluation process in teacher education programs is further emphasized by NCATE in the publication *Standards for Accreditation of Teacher Education* (1981).

Maintenance of acceptable teacher education programs demands a continuous process of evaluation of the graduates of existing programs, modification of existing programs, and long-range planning. The faculty and administrators in teacher education evaluate the results of their programs, not only through the assessment of graduates but also by seeking reactions from persons involved with the certification, employment, and supervision of its graduates. The findings of such evaluation are used in program modifications. (pp. 10–11)

The accreditation process for a technology education program usually is comprised of the following elements: (1) a written report (self-study) prepared by the program faculty and/or administration documenting the claim that it effectively meets the standards of an effective program; and (2) a team of trained evaluators reviews the written report and votes to accredit or not to accredit the program, based upon available documentation. The results are forwarded to NCATE for inclusion in the institution report. It should be noted that NCATE classifies technology education as a special program. This means that there is not a visitation as part of the accreditation process. This method of program evaluation rarely involves teacher educators or their students, except in the role of those being evaluated.

Examination of Teacher Effects. According to Borich (1979), this approach evaluates a teacher education program in terms of the effects its graduates have on the learning of public school students. This approach makes use of the ultimate teacher education program criterion, teacher efficiency, as measured by pupil learning. The procedure for implementing this approach is to ask teachers to teach identical units to pupils and through the administration of a pretest and posttest determine the learning comprehension of the students. It is obvious that the logistics required to carry out this plan are difficult. The results of this approach would not be useful for making program improvements because the results deal with the effectiveness of the overall program and not separate distinct elements.

SUMMARY

This chapter examined the major issues related to pre-service teacher preparation programs for teachers of construction in technology education. The design and implementation of a program should begin with a vision. This vision should reflect two guideposts. First, the program should be guided by the attributes of an effective teacher of construction in technology education. The second guidepost is a constant awareness of the major criticisms of traditional programs. Specific qualities and general cognitive abilities of potential teachers were identified and discussed. In doing this, some of the central issues involved in teacher selection were also presented. The major components of a model teacher education program were identified and examples of appropriate learning experiences for each component were presented. Professional technology teacher educators must continue to reflect on this important matter.

REFERENCES

Andrew, M.D. (1989). Subject-field depth and professional preparation: New Hampshire teacher education program. In J.L. DeVitis & P.A. Sola (Eds.), *Building bridges for educational reform* (pp. 44–62). Ames, IA: Iowa State University Press.

Bonstingl, John J. (1992). *Schools of quality: An introduction to total quality management in education.* Alexandria, VA: Association for Supervision and Curriculum Development.

Borich, G.D. (1979). *Three models for conducting follow-up studies of teacher education and training.* (Report of the OECD on the evaluation of in-service education and training). Paris, France: OESD.

Brandt, J. (1991). On teacher education: A conversation with John Goodlad. *Educational Leadership. 49*(3) 11–13.

Bush, R. (1960). *The liberal and technical in teacher education.* New York: Teachers College Press.

Clark, D.L. (1986). Transforming the structure for the professional preparation of teachers. In J.D. Raths, & L.G. Katz (Eds.), *Advances in teacher education: Volume 2* (pp. 1–18). Norwood, NJ: Sablex.

Custer, R.L. (1990). Liberal education and the practical arts. *Journal of Industrial Teacher Education. 27*(4) 46–56.

Devier, D. (1981). Industrial arts teacher education student recruitment practices and their effectiveness in the state of Ohio. *Dissertation Abstracts International,* 42/05A, 2001A.

Devier, D. (1986). *Industrial arts/technology education recruitment.* Unpublished raw data.

Devier, D. (1987). The best and brightest to teach technology. *The Journal of Epsilon Pi Tau. 13*(2) 28–31.

DeVore, P.W. (1975). *Technology - Its impact on industrial arts education.* Blacksburg, VA: Virginia Polytechnic Institute and State University. (ERIC Document Reproduction Service No. ED 118–750).

Heath, P. (1984). An examination of the curriculum of early field experiences in selected teacher education programs in Ohio. *Dissertation Abstracts International,* 45/06A, 1724A.

Henak, R. (Ed.). (1991). *Elements and structure for a model undergraduate technology teacher education program.* (CTTE Monograph 11). Reston, VA: International Technology Education Association.

Householder, D. (1992). Redesign of technology teacher education: Model programs for the future. *Camelback Symposium: A compilation of papers,* (pp. 4–9), Reston, VA: International Technology Education Association.

Howey K.R., & Strom, S.M. (1987). Teacher selection reconsidered. In M. Haberman, J.M. Back, & J.M. Norwood (Eds.), *Advances in teacher education: Volume 3* (pp. 1–30). Norwood, NJ: Sablex.

International Technology Education Association & Council on Technology Teacher Education. (1989). *NCATE folio preparation handbook for technology education undergraduate programs.* Reston, VA: Author.

Jacobson, R.L. (1986). Tomorrow's teachers: A report of the Holmes Group. *The Chronicle of Higher Education. 32*(6) 27–32.

Katz L., Raths, J., Mohanty, C., Kurachi, A., & Irving, J. (1981). Follow-up studies: Are they worth the trouble? *Journal of Teacher Education. 32*(2), 18–24.

Lacroix, W.J. (1987). Characteristics for successful professional performance as a technology education teacher. *The Journal of Epsilon Pi Tau. 13*(2), 32–39.

National Commission for Excellence in Teacher Education. (1983). *A nation at risk.* Washington, DC: U.S. Government Printing Office.

National Commission for Excellence in Teacher Education. (1985). *A call for change in teacher education.* Washington, DC: U.S. American Association of Colleges for Teacher Education.

National Council for Accreditation of Teacher Education. (1981). *Standards for accreditation of teacher education. The accreditation of basic and advanced preparation programs for professional school personnel.* Washington, DC: Author.

Petrie, H.G. (1987). *Teacher education, the liberal arts, and extended preparation programs.* Albany, N.Y: Nelson A. Rockefeller Institutes of Government.

Secretary's Commission of Achieving Necessary Skills. (1992). *Learning a living: A blueprint for high performance.* Washington, DC: U.S. Department of Labor, U.S. Government Printing Office.

Stufflebeam, D.L. (1982). Exploration in the evaluation of teacher education. In S.M. Hord, T.V. Savage, & L.J. Bethel (Eds.), *Toward usable strategies for teacher education program evaluation* (pp. 132–144). Austin, TX: Research and Development for Teacher Education.

Weins, A.E. (1987). Technology education and the liberal arts. *The Journal of Epsilon Pi Tau. 13*(2), 16–23.

Wescott, J.W. (1989). The relationship between Scholastic Aptitude Test scores and the academic success of industrial art/technology education majors. *The Journal of Epsilon Pi Tau. 15*(1), 32–36.

Wright, R.T. (1990). Changes in curriculum, students, and educator's role call for new competencies. *School Shop/Tech Directions, 50*(1), 25–27.

8

Instructional Approaches To Teaching Construction In Technology Education

James E. LaPorte, Ph.D.
Technology Education Program
Virginia Polytechnic Institute, Blacksburg, VA

Approaches to teaching technology education have played a rather different and interesting role relative to other subject areas or disciplines in the school. With virtually every other subject in the school, the content to be learned is first determined — usually what is contained between the covers of the textbook is selected. Then, appropriate approaches and methods to deliver the content are developed.

In technology education, the process is often reversed. Perhaps due to the fact that activity-based instruction has been the hallmark of the field since its inception and the fact that written instructional materials (i.e., textbooks) are relatively new to the field (compared to other disciplines), instructional planning often starts with the activity (or project). Then, the actual learning outcomes are "reverse engineered" from the activities. In other words, a course is planned by selecting the projects and activities in which the students will be engaged and then they are analyzed to determine what the students will learn on an *ex post facto* basis.

Such an approach is deplorable in light of virtually all the wisdom of authorities on the curriculum and course development process. It is what Bensen (1980, p. 12) called the "seat of the pants approach." With more curricula written and available in published form, along with a marked increase in textbooks, especially at the middle school level, this backwards approach has diminished and will no doubt continue to do so in the future. Yet there is evidence that this practice is continuing based on the diligence with which teachers are collecting design briefs for problem-solving activities that this writer has observed.

But perhaps there is a bit of rationality in why technology educators seem to be backwards. For other disciplines, the method used to teach a particular instructional unit represents what students and teachers will do to facilitate learning. Learning starts with the content which is then delivered by the doing. On the other hand, doing is the essence of technology and is the principal element in most scholarly definitions of the term. In some respects, the doing is both the content and the method of technology education. It distinguishes, to a large extent, the study about technology that is a part of many other disciplines (e.g., natural and social sciences, mathematics) from the doing technology that has uniquely characterized technology education programs. Such a perspective might at least explain why technology educators may use a non-traditional, albeit backwards, approach to developing course and curricula.

The basic tenets of the philosophy of technology education are much broader in scope than those that guided industrial arts. As mentioned earlier, an abundance of textbooks have been published exclusively for technology education which reflect this breadth. In the construction area, textbooks are no longer one and the same with trade-based vocational courses and are much more inclusive of construction *in toto* and more appropriate for general education purposes.

Other disciplines have been envious of the hands-on approach used in technology education and have sought for years to make courses more applied and filled with meaningful activities that truly engage students. Most of the recent reform initiatives in science and math such as Project 2061 (American Association for the Advancement of Science, 1989) and the standards for mathematics developed by the National Council of Teachers of Mathematics (1991) have emphasized the need for hands-on learning.

It is a difficult challenge taking basic elements of a discipline and developing meaningful, engaging, and motivating learning activities that correlate with those elements. Think of the challenge for the social studies teacher in developing such activities in history. Likewise for the science teacher in atomic theory. At least intuitively, this is a much more difficult problem than structuring the content of the discipline. And, the more abstract the concepts, the more difficult the challenge.

When industrial arts was material-based (woodworking, metalworking, etc.), the methods used seemed to evolve from the content rather easily. This was especially true when the main objective was to develop tool skills. But now, with content structures evolving that are more inclusive of the totality of technology (if, in fact, that is achievable), technology education must meet the same challenges as other disciplines in developing approaches to teaching. There are aspects of technology that are quite abstract. And the scale of certain technological artifacts and practices is such that it is virtually impossible to portray them validly in the school setting. Furthermore, the down-scaling of technology so much results in artificial, contrived, and meaningless experiences for students. Pucel (1992) stated the dilemma quite clearly: "People do not build bridges with straws, tape, and scissors. They build bridges with cement, steel, and hammers" (p. 30).

So how are we meeting the challenge of developing effective teaching approaches for the ever-expanding and increasingly abstract content of technology? Back to the textbooks. The new technology education textbooks are much more attractive than their predecessors with far more illustrations and the generous use of color. And they certainly include up-to-date topics such as computer-aided design, lasers, robotics, and so forth. But several examples can be found where the breadth is so extensive that there is no substance — nothing that the learners can "sink their teeth into." Some texts appear more like a series of late-breaking news stories on technological developments than as a written aide to learning the discipline of technology.

But even more serious is how some of the new texts either avoid the issue of meaningful learning activities and corresponding teaching approaches, or they resort to those that were used in the past when the scope of the content was much narrower. When one reviews some of the articles written in journals in the field since the evolution to technology education, ample examples of modern topics in virtually all of the vast areas of technology can be found. But more the rule than the exception, the activities suggested are "discuss (this topic) with other students in the class" or "go to the library and research (this topic)." It appears that our ability to conceptualize the content of technology education has out-stripped our ability to develop activities that involve meaningful doing. We need to catch up lest we become just like our counterparts in other disciplines, relying exclusively on lecture and discussion and losing our laboratories whereby we can provide students with the real, practical, hands-on experiences of which others are envious.

The foregoing sets the stage for what is to follow. Though in some respects it reads like a conclusion to the chapter, the writer was compelled to bring to the surface some of the issues that bear upon teaching approaches to construction, thereby setting the context and providing the reader with a more critical and analytical perspective.

THE CONTENT VERSUS PROCESS ISSUE

The issue of content versus process has been ongoing in educational deliberations, especially in recent years (e.g., Riner, 1990). The dichotomy has also been an ongoing issue in technology education as evidenced by the fact that the 1992 Technology Symposium held at Millersville University

(McCade, 1992) was devoted to this issue. The issue has implications for both content and teaching strategies.

The content side of the dichotomy holds that there is a body of knowledge that all members of a society should know and be able to apply. The body of knowledge selected is based on cultural, scientific, or utilitarian criteria. Typically, a group of experts representing the discipline put their collective thoughts together. In fact, some hold that such a codified structure of content is one of the hallmarks of a discipline.

The process side of the dichotomy holds that the education system must provide the means for students to "know how knowledge can be found, how it is created, and how to transform knowledge into utilitarian solutions that confront individuals" (Riner, 1990, p. 63). While not denying the importance of content mastery, the process-oriented educator holds that the learner will obtain the skills to seek out the knowledge of content that is needed to solve the problems with which he or she is confronted.

The content educator is committed to the idea that there is a body of knowledge that must be transmitted to students. Lesson plans reflect the sequential presentation and learning of this content. The process educator does not see the importance of a prescribed body of knowledge. If students can be equipped with the skills to seek out the knowledge that is pertinent to their interests and the problems, and challenges they will meet, this is more lasting and valuable than learning a fixed body of content that will inevitably become obsolete and even more difficult to recall.

Thus, the content versus process issue is rooted in the beliefs that the teacher has about how students learn, his or her perceived role as a teacher, and their experience. These beliefs will determine, to a large extent, the approaches that predominate one's teaching style.

Early industrial arts construction programs were very content oriented. Likely this was due in large part to the vocational influence that permeated the field at the time. With an emphasis on tool skills, the instructional strategies used often paralleled trade-based programs and often differed only in the amount of time the students had to practice their skills. The analysis of trades and jobs related to construction determined not only the content, but the methods by which the content was delivered. Some feel, as Lux does, that the trade and job analysis influence hampered curriculum development and methodology, calling it the "scourge of industrial arts" (Lux, 1979).

The Industrial Arts Curriculum Project (IACP) developed a structured body of knowledge for construction practices (Towers, Lux, & Ray, 1966) and then developed a program to teach that structure at the junior high level through a program called the World of Construction (Lux & Ray, 1971). This was a dramatic departure from the earlier trade-based programs and their emphasis on tool skill development. Also, the content structure was very comprehensive, adding management and personnel practices as well as those dealing with production. However, the two major activities in the program, the Dream House and the Construction Module, primarily emphasized residential construction. Thus, IACP was content oriented, but the strategies moved away from trade-skill development with an emphasis on conceptual learning.

As the field changed to a technology base rather than an industrial base, problem solving became pre-eminently emphasized as both content (the problem-solving method, ala the scientific method) and process (solving practical problems). In the evolution of the field, embracing problem solving perhaps offers the first real opportunity to teach technology as a process. That is, the process of problem solving could become the principal goal of technology programs. In the ideals of the process educator, students will be able to learn all the content that they need to know through the problem-solving process. On the other hand, content-oriented educators might subscribe to the problem-solving process, but only to the extent that it is one of several means to teach the content.

Perhaps the dichotomy between content and process is inappropriate. It could be argued, as Riner (1990) did, that content and process are both essential elements of rational curriculum development and, in fact, are mutually dependent. Brandt (1988) noted that most educators do not draw such a dichotomy. Only time will reveal whether technology education continues to be content-based or whether it moves toward an emphasis on process. Likely, compromise will prevail—the apparent consensus at the Millersville Technology Symposium. Because of its narrow, trade-based heritage, perhaps construction has a longer way to go than some of the other areas of technology education in moving to a more process-oriented program.

LEARNING STYLES

People learn in different ways. The preferred and most effective way in which an individual learns has become known as their learning style. The concept of learning style is not a new idea. Claxton and Murrell (1984, p. 3) provided evidence that differentiation in learning style can be traced back at least 2500 years. Even teachers with a minimum of experience soon recognize that students have different styles of learning. What is important in recent years is that the existence of learning styles and their categorization has captured the efforts of researchers. Guild and Garger (1985) stated that the realization of learning styles is "the most important concept to demand attention in education in many years" (p. viii).

Of equal importance is the discovery that teachers tend to predominantly teach in a style that is consistent with their own learning style (e.g., Witkin, 1976). Though the research is not conclusive, it is nonetheless significant, as well as intuitively logical. Humans tend to generalize their own experience to the situations that others face. That is, "If this is the way that I learned best, then it must be best for my students."

One of the most predominate people in learning style theory is David A. Kolb. He has developed a model of learning styles and developed an instrument to measure one's predominant learning style. In addition, he has conducted a considerable amount of research to validate his ideas.

Kolb (1984) proposed that learning occurs through a four-step process. First, the learners have a concrete experience in which they are fully involved. Second, the concrete experience is followed by a period of reflective observations in which they look at their experiences from varying perspectives. Third, they fit the experiences into their mental schema by developing generalizations or principles and then incorporating them into guides for further action through a process called abstract conceptualization. Finally, the learner tests the conceptualized ideas in new situations through active experimentation.

Kolb categorizes learners according to how experiences are grasped and then how the experiences are transformed or processed. Some prefer to take in information in concrete ways while others prefer more abstract methods. In processing information, some rely upon reflective observations while others change the information through active experimentation so that it fits their personal mental schema.

From the ways in which people prefer to take in information and the ways in which they process information, Kolb hypothesized four different learning styles. Divergers take in information through concrete experience and process it through reflective observation. Assimilators take in information through abstract conceptualization and process it through reflective observation. Convergers take in information through abstract conceptualization and process it through active experimentation. Accommodators take in information through concrete experience and process it through active experimentation.

Kolb (1981) assessed the learning styles of a group of 800 managers and graduate students to determine their learning style in relationship to their chosen major in college. The results of his study are superimposed upon his learning style model in Figure 8–1.

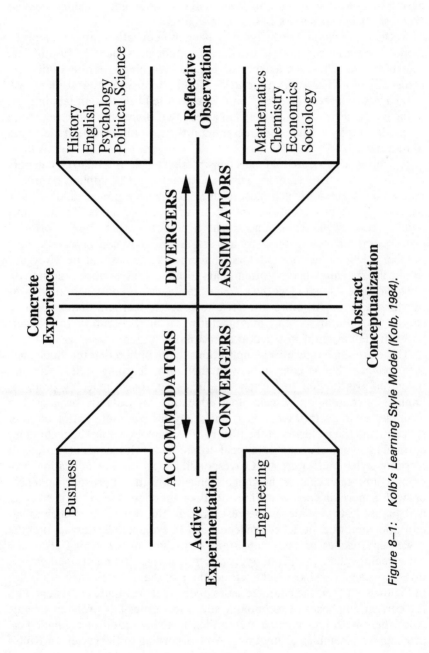

Figure 8–1: Kolb's Learning Style Model (Kolb, 1984).

Henak (1992) applied Kolb's work to technology education, providing a description of the four learning styles and corresponding teaching styles. A synthesis of Henak's work is shown in Figure 8–2.

In theory, technology education is a general education program and the learning experiences that it provides to students should be useful and valuable to all. Thus, the teacher should plan learning experiences that cut across all four of Kolb's learning styles. This better assures that students will not only learn more effectively, but they are more likely to maintain interest in the course as well. It also helps to assure that there is not discrimination by gender nor by ethnic origin, essential if the general education ideals are to be realized.

It is fairly safe to state that most construction teachers tend to be concrete in their approach to teaching (no pun intended). The applied nature of industrial arts, under which most teachers today were prepared, deals with concrete ideas by definition. Industrial arts tended to be fairly tightly structured as well. Thus, one might hypothesize that the majority of experienced technology teachers are assimilators and convergers. A small number might be accommodators. It seems like few would be divergers. Could this explain why one often hears teachers lament about the narrow range of students that elect the courses they teach? Is construction, the way in which it is taught, more attractive to concrete learners than communication, transportation, and manufacturing? Could attention to learning styles in the design of instruction attract more students?

There are some interesting implications to the notion that teachers teach in a manner that is consistent with their own learning style. Whereas, industrial arts tended to be concrete in its approach, teaching practices implied by contemporary technology education curricula are broader. For example, the impacts of technology on society, the overt integration of other subjects, and the inclusion of the history of technology within the technology curriculum would seemingly appeal to students with a wider variety of learning styles and interests. Correspondingly, as teacher education programs continue to change they might conceivably attract prospective teachers with a more diverse set of interests than has been true in the past. And, is it plausible that some of the resistance to the transition to technology education among experienced teachers can be explained by the fact that the newer approaches are not consistent with their own learning style and consequently, their preferred approach to teaching? As mentioned earlier, divergers seem least like teachers trained under the industrial arts era (refer to Figure 8–2). Yet, the characteristics of divergers are quite consistent with the current emphases in technology education, especially problem solving. And when math and science are applied to the solution of technological problems, it resembles engineering. And, according to the results of Kolb's

	Learning Style	Teaching Style
Divergers	Strengths in imaginative ability, seeing things from different perspectives, and then putting them into a meaningful whole. Good at brainstorming. People-oriented. Emotional.	Avoid conventions and rules. Encourage development of creative and critical thinking. Variety of activities and exposure to different people. Evaluation considers students' feelings.
Assimilators	Systematically research ideas, theories, and processes; and then, transform them by describing, conceptualizing, generalizing, and diagramming. More interested in logical soundness of an idea than practical application.	Instruction is usually teacher-centered. Lecture is a common method. Intellectual pursuits are stressed. Classes are highly structured with emphasis on logical reasoning and analysis. Detailed, comprehensive grading system is used.
Convergers	Like practical application of ideas and theories. Grasp ideas through careful and systematic study and then transform by testing, trying, applying them to practical situations. Move quickly to find answers. Good at deductive reasoning, problem solving, and making decisions. Unemotional. Prefer things over people.	Orderly classroom and schedule is followed. Emphasis on practical learning and concrete experiences. Conventional processes are used. Teacher-centered approach.
Accommodators	Grasp information by engaging in testing, trying and using. Transforming is done by trial and error and observation.	Expect students to relate to the real world, express themselves, and respond to current issues and problems. Loose flexible schedule. Evaluation emphasizes practical application.

Figure 8–2:   *Descriptions of learning styles and teaching styles (Adapted from Henak, 1992, pp. 24–25).*

work summarized in Figure 8–1, most engineering majors tend to be divergers. A concerted research effort is needed in this area.

SELECTED APPROACHES FOR TEACHING CONSTRUCTION

Approaches to teaching a subject like construction can be organized and presented in a variety of ways. They could, for example, be organized as dichotomies such as child-centered versus subject-centered, field-dependent versus field-independent, and lecture versus discussion. They could also be presented categorically, referencing learning styles, for example. Teaching approaches can also be presented with respect to levels, ranging from microscopic aspects of teaching (e.g., questioning techniques) to global. The presentation used here is global with the reader given the freedom to dichotomize and categorize independently. But, the following points should be kept in mind as the reader reviews the approaches presented:

1. The approaches are certainly not totally inclusive of all possible approaches nor are they pure examples of any particular approach.

2. The approaches are not mutually exclusive. Elements of one approach can be found in others.

3. There is no single correct approach prescribed. It goes without saying that variety in teaching approaches is essential and substantiated by abundant research.

4. The approaches are not listed in any particular order.

The Enterprise Approach

The Enterprise Approach, as an instructional strategy for construction, is parallel to the enterprise approach used in manufacturing. The context is that of entrepreneurship in which risks are taken in hopes of making a profit. The enterprise approach was a key instructional strategy in the Industrial Arts Curriculum Project (IACP) World of Construction program and remains important in construction textbooks which followed IACP (e.g., Henak, 1993). The content is usually organized into the broad areas of management and production (or management, personnel, and production, as in the IACP program). The instructional strategies grow out of the content itself.

Consistent with entrepreneurship, students are organized into teams, representing a construction company. Various construction projects are specified to the students and these projects become the core of the learning experience. The students learn various techniques for building structures, as they perform them through the project. This constitutes the production portion of the content. They also learn management practices through planning, estimating, and bidding on the project as well as organizing and controlling the project as it is constructed.

An example of the enterprise approach is in the construction of storage buildings. The teacher might provide a set of drawings and specifications for one or more storage building designs to the students. From these drawings, each team prepares and submits a bid for the project, including both material and labor costs. Bids are then awarded to the lowest bidders and construction of the project begins. Members of each team organize themselves and establish a management system so that each person has prescribed responsibilities. A work schedule is instituted. Careful accounting of labor and materials is important so students learn the economic and management principles essential to the approach. When the structure is complete, a determination of the profitability of the enterprise is made.

The enterprise approach is an excellent method to teach the students about the construction industry from the viewpoint of a consumer as well as a potential career. Thus, it meets general education as well as prevocational objectives. The team approach is consistent with many of the recommendations for educational reform (e.g., The Secretary's Commission of Achieving Necessary Skills, 1991). As with the enterprise approach in manufacturing, the sense of realism, importance, and urgency in completing the project creates a motivational learning environment. Since the project is real, the students are taken several steps beyond simple role playing.

One of the problems with the enterprise approach is that it is difficult to engage students in projects that are not principally residential construction. Moreover, storage buildings are not usually constructed in the same way as even homes are. Thus, the students are confined to storage building construction rather than even the delimiting aspects of residential construction. At its worst, this approach limits learning experiences to a few trades, resembling a sort of miniature vocational experience. If the management element is eliminated, as often seems to be the case, then it certainly is reduced to a vocational experience, albeit of very questionable value for any educational purpose.

Another problem with the enterprise approach is the expense of the materials. Storage buildings (and comparable products produced using the enterprise approach) are saleable products. The proceeds from their sale should at least cover the cost of materials. However, the burden for selling often falls upon the teacher. This can be time-consuming and frustrating and results in low acceptance by many teachers. What's more, the market for storage buildings can often be saturated in a short period of time, especially in smaller communities.

The problem of the limited experiences that the storage buildings provide can be solved. Instead of applying the typical practices of constructing a storage building, the practices used to construct an actual dwelling could be used. A concrete block foundation could be laid using excessive lime in the mortar so that the blocks can be easily recycled. The floor-frame, walls, and roofing system could be done just as they are in a house. An electrical-service entrance, plumbing system, and duct work for a heating system could be installed. Standard windows and doors could be used. Floor tile could be installed along with gypsum board on the walls and ceilings. Using this approach, the students are exposed to a wide variety of construction experiences and the potential for teaching management concepts is likewise enhanced. But using the residential-construction practices adds to the expense of the structure.

The solution to the expense problem, as well as the problem of the burden placed upon the teacher, is to work with a local building materials dealer. Many such dealers market storage buildings for the homeowner. They may be willing to supply the materials, including the excess materials required to provide a broad learning experience for the students, in exchange for the finished storage building. In addition, building material dealers usually have the equipment to pick up the completed storage building at the school and transport it to the place of business, thus taking the marketing burden from the teacher. Even with the excess materials, the completed storage structure usually provides a profit margin for the dealer that is competitive with those that the supplier purchases from wholesalers.

The enterprise approach can be enhanced by having students involved in the design of the structure. One way to do this is to have teams of students compete to design the best structure and then use that design among all the teams. The management aspect of the enterprise approach offers great, real-world potential for the integration of computers. Computer-assisted design can be used during the planning and design process. Spreadsheets can be used to prepare estimates and bids, as well as keeping track of labor and materials during the course of the project. A database system can be used to track inventory.

The Interdisciplinary Approach

The interdisciplinary approach revolves around a central theme or project that involves all, or nearly all, disciplines in the school. The central-theme idea has been used in the past. An example is Earth Day which was initiated in the 1960s. In its ideal form, Earth Day tied all the disciplines together with the central theme of environmental issues and action. Social studies students focused on the relationship between the environment and the quality of life, science students studied air pollution, and technology students solved problems related to recycling. In some schools, the culmination of Earth Day was a comprehensive environmental improvement effort on the school campus or at some location within the community.

Interdisciplinary approaches have had their fits-and-starts for nearly the course of educational history. They have a good foothold at the elementary level (perhaps due to the relative ease with which they can be implemented with a single teacher) and they are gaining significant ground at the middle school level. The *Middle School Journal* is replete with articles describing interdisciplinary efforts.

The movement toward the interdisciplinary approach is based on a rather simple, but logical, rationale. Subjects and disciplines cannot be effectively studied in isolation from one another. Rather, there is a symbiotic relationship among them. Needless to say, schools are structured with artificial (as well as physical) boundaries around disciplines. And, there is a continuum of the width of the boundaries as one moves from elementary school (thin) to the collegiate level (thick). Using a core project or theme, the students see, understand, and appreciate the symbiotic relationship.

An example of the interdisciplinary approach that involves construction is the Space Simulation Project, pioneered by two public school teachers in a suburb of Houston, TX (Bernhardt & McHaney, 1992). The goal of the project is to simulate a space-shuttle mission. A simulated space shuttle is constructed in which students will spend several "real" days playing the role of astronauts. All life-support systems are included within the shuttle, including food, air supply, waste disposal, and so forth. Experiments are conducted during the course of the mission. As in the real world, a mission-control center is established. A video camera at the control center and within the shuttle provide two-way communication. Computers are used to track and log the progress of the mission as well as for a multitude of other purposes.

Since the project is technological in nature, the technology program might play a coordinating role. The role of the construction class might be to construct the simulated shuttle and mission-control center. However, each subject area in the school must play an important part in order for the approach to work. The key is cooperation, not competition. Business students might keep track of and analyze the economic aspects of the mission. Science students might develop the experiments to be conducted during the mission. Math students might calculate rocket trajectories and provide course information. Art students might be involved in designing a logo for the mission. Social studies students might deal with the social issues of life on another planet or the strategies to get along with others in a confined environment. Health classes might study the results of zero-gravity conditions on the human body. Physical education might deal with exercise routines for the crew to keep fit.

As with any interdisciplinary approach, there are some trade-offs. One is the traditional subject matter "turfdomship" that has been a significant impediment to interdisciplinary approaches for years. Participating teachers must relinquish the traditional boundaries that they have marked out for their disciplines and adopt the attitude that "as long as students are learning, it makes little difference how it is occurring or who is the principal facilitator." Even within technology education, there are turf battles. For example, should the mission-control center be the primary responsibility of the construction class or the communication class or the manufacturing class? Or should the whole project be mainly the responsibility of the transportation class since, after all, the space shuttle is primarily a transportation endeavor? Sometimes, perhaps, the main beneficiaries of the interdisciplinary approach are the teachers when they discover how artificial and ridiculous are the boundaries that separate disciplines and sub-disciplines.

Another trade-off of the interdisciplinary approach is the compromise in the content covered. An interdisciplinary approach, such as the space-shuttle mission, takes instructional time. Consequently, a portion of what is normally covered in a construction course, for example, must be replaced by the content that evolves from the interdisciplinary approach. Teachers usually believe in the learning experiences they have designed into their courses. It is natural that they will question the value of replacing some of those learning experiences.

A third trade-off is also a dilemma of content. An interdisciplinary approach may engage students in learning experiences that have a questionable relationship to the course. In the space-shuttle mission, for example, the type of construction that the students do in constructing the simulated space shuttle may have little relationship to construction in the real world. In fact, the experiences may be more similar to stagecraft than to any real construction practices. This trade-off is particularly disconcerting for construction teachers who place an emphasis in their program on construction skills. But, a creative teacher can relate the space-shuttle experience to the previous content of the course. Besides, the construction practices used to build an actual structure in space bear little semblance to those used on earth. Certainly, management and personnel practices that relate directly to construction can easily be integrated into most interdisciplinary approaches. All things considered, the interdisciplinary approach may be most consistent with the process-oriented teacher.

The Integrative Approach

The integrative approach bears some resemblance to the interdisciplinary approach in the sense that there is coordination among two or more disciplines. However, it does not typically focus upon a core project. The basic idea of the integrative approach is to recognize that certain content taught in one class parallels the content in another class. At its basic level, the integrative approach simply sets about identifying the common content in the courses and then the parallel units are taught at the same time in both classes. For example, if students in construction are engaged in a unit in contract estimating, this unit might be timed so that it occurs at the same time as a unit in calculating areas and volumes is taught in the math course. Or a unit on heating and insulation in construction might be taught at the same time as a unit on thermal energy in science class.

A second level of integration attempts to include the teaching, or review, of the content from other disciplines that has a direct relationship to what is being taught in technology education. For example, the theory of heat transfer (from science) is taught as a part of unit on heating systems in a construction class. Or, the social implications of housing deficiencies in third-world countries (social science) is taught as part of a housing design activity. Or, the techniques of calculating areas and volumes (mathematics) is taught by the construction teacher as part of a unit on conducting an energy audit for a dwelling. This is the integration approach used by the Appalachian Technology Education Consortium (ATEC), headquartered at West Virginia University.

A third level of integration identifies the common content in two or more classes and then the teachers involved cooperatively plan their instruction so that it is delivered in a coordinated manner. Though a requisite, this type of integration goes far beyond simply sequencing content so that it occurs concurrently. It requires common planning time and a willingness to devote the extra effort that is necessary to make it work. In essence, this third level of integration is the same as the second level except that the common content is taught by the teachers of the respective disciplines involved. The various disciplines become referents for one another. If students need to know how to solve equations with one unknown before they can solve a problem in determining the R-value of insulation, the necessary mathematics skills are taught (or reviewed) by the mathematics teacher. If students need to know the theory of how heat is transferred in order to solve a problem involving insulation, that theory is taught by the science teacher.

This third level of integration is what is used in the Technology/Science/ Mathematics Integration Project (LaPorte & Sanders, 1992). The idea behind the project is to confront the students with a technological problem that requires the application of science and mathematics principles for its effective solution. The underlying science and mathematics concepts are taught by the science and math teachers. After the solutions have been developed, they serve as practical examples of the application of the principles. An example in construction is the composite beam. Students are challenged to design the shape of a concrete beam that fits into a prescribed volume. The beam is reinforced with specified recycled materials. The students compete to determine which beam is most efficient (weight that the beam supported divided by the weight of the beam).

The fourth level of integration involves a total team-teaching approach. Students are "block scheduled" for a period of time (e.g., three class periods). The teachers involved (e.g., technology, math, and science) work with the same group of students during the block of time. Working as a team, the teachers plan the instruction for the entire year. They make decisions on who will teach what and when, which activities will be used, and how students will be evaluated.

As one moves from lower to higher levels of integration, the resources involved increase. Higher levels of integration require more planning time, a common planning time, more commitment from the teachers and the administration (especially), and a revision of the school schedule (a "sacred cow" to many administrators). More importantly, as one moves up the levels of integration, a greater compatibility of teacher personalities is required. This compatibility is perhaps the greatest impediment to higher levels of integration. And, if one wants to be pessimistic, a look at the failure rate of small-business partnerships (or marriages, for that matter) offer support. Yet, in theory, the integrative approach (as well as the interdisciplinary approach) are, hands-down, the most promising.

The Research-Project Approach

In the research-project approach, students are engaged in conducting actual technical research. The research problem might be prescribed by the teacher and directly related to the content of the course. For example, if one of the content units is Thermal and Sound Insulation, then teams of students might be challenged to develop wall panels of prescribed dimensions that provide the maximum resistance to sound and heat. The panels would then be tested using testing apparatus prepared by the teacher or purchased by the school. Data are collected and analyzed and conclusions drawn from the findings.

Another way in which the research-project approach might be implemented is to allow the students to formulate the problem to be investigated. In this way, the problem is directly related to the interest of the student-researcher. For example, a team of two students might be interested in the earthquake resistance of various highrise-building designs. Once the problem has been stated and articulated, the students would then prepare models of the various designs that are to be tested. Concurrently, they would design and build an apparatus to simulate earthquake conditions by which the models can be tested. Like the prescribed research project, conclusions are drawn from the findings obtained from the collected data.

Regardless of which approach is taken, a research report is a necessary outcome. The report should include the standard sections found in most research reports. The first section should include a statement of the problem. The second section includes a review of literature that pertains to the problem. In the third section, the methods that are to be used in the research are described in detail. If apparatus is to be built, such as that mentioned in the above example, it would be described in this section. Drawings of the apparatus would also be included. The fourth section reports the findings of the study; that is, the results of the testing. The last section of the research report includes the conclusions reached through the study.

The concepts of validity, reliability, and generalizability are essential to the research-project approach. Students must understand threats to the consistency and trueness of their data and recognize the extent to which they can be legitimately generalized to other situations. This approach is an excellent vehicle to accomplish this and has great potential to educate students to be wise consumers of research for general education purposes. Thereby, they can more critically evaluate what is written in popular magazines such as *Consumer Reports, Popular Mechanics, Popular Science,* and *Motor Trend.*

The research-project approach has many advantages. If the problem to be investigated grows out of the students' own interests, motivation will most certainly be enhanced. Individual differences can be readily addressed through the complexity of the problem chosen, the thoroughness with which it is investigated, and the relative capability of each of the team members. There are excellent opportunities for the application of both science and mathematics. This approach has particularly rich opportunities for physics students since technical construction problems often directly apply physics principles, albeit, "doing engineering." Mathematics applications can range from simply calculating the averages of the results of tests, to sophisticated regression analyses in which predictions are made. Again, individual differences are easily addressed.

Getting students to write in the technology class is a complaint often heard from teachers, especially those who deal primarily with students that do not have a strong interest in writing to begin with. However, if the motivation to do the research project is truly intrinsic, then the motivation to share the results with others will often follow. The practical nature of the research project may give meaning to the need for writing in comparison to the contrived contexts which students often confront. As the research project offers excellent means to integrate science and mathematics, the research report leads to a natural integration of language arts. Thus, the research-project approach could be easily extended into a true, multidisciplinary-learning experience.

The research-project approach also provides good opportunities to utilize the capabilities of computers. Spreadsheets can be used to collect and analyze data. Charts and graphs can be imported into the word-processed reports. Presentation graphics software can be used to enhance the presentation, either through the generation of slides or overhead transparencies or through using the computer directly to present a slide show.

This approach can be a good public relations vehicle for the construction program in the school. This is especially true for parents of students bound for engineering and technical careers. It can also serve well to establish the credibility of the construction program in the minds of practicing engineers and related groups. The finished projects can make attractive exhibits at science fairs. As Lux (1977) noted, many science fairs are really technology fairs. Or, the projects could become the centerpiece for a true technology fair. Local newspapers might publish the results found from some of the more interesting projects.

A disadvantage of the research-project approach is time. If the student chooses the problem to be investigated and designs and develops the necessary testing apparatus, a large proportion of the school term can be consumed. The time requirements can be dramatically reduced if the teacher prescribes the problem, but the ownership that the student has to the problem and the resultant motivation will no doubt be reduced. On the other hand, the competition that occurs in trying to see who can develop the best solution to a common problem may increase motivation for some students.

The student-selected research project can also create a burden for the teacher in helping the students formulate the problem, design and build the testing apparatus, and prepare a written report. The affinity that the teacher has toward using this approach is dependent on their willingness to accept this burden. It is also related to the teacher's philosophy regarding content versus process. The research-project approach is primarily process-based.

The Systems Approach

The systems approach has its roots in the work of Bertalanffy (1928). During the 1950s, systems thinking began to receive widespread attention among scholars, especially those in engineering and mathematics. The journal, *General Systems*, appeared in 1956. Within technology education attention to systems theory increased with the *Jackson's Mill Curriculum Theory* (Snyder and Hales, 1981) in which technology was described as a human adaptive system. By the mid-1980s, several textbooks had been published which used the word "system" in their title and were organized, in varying degrees, around systems theory. The recent curriculum effort led by Savage and Sterry (1991) utilizes an adoption of systems theory extensively. Historians of technology education may one day refer to the decade of 1985–1995 as the "systems era" in technology education.

The underlying concept of systems theory is based upon the generalized systems model, as shown in Figure 8–3. The generalized systems model is, in essence, a particular way of thinking about something and may be applied to virtually anything. The heart of the system is the processing. Processing requires input and results in output. With the three elements of input-process-output, an open-loop system exists. If feedback from the output to the input is added, the system is referred to as a closed-loop system.

The methodological aspects of systems theory in technology education are closely tied to the content aspects and it is difficult to separate them. In many respects, the content drives the method. In construction, the systems approach has been a springboard in thinking to get away from obsolete approaches that result in an exclusive emphasis on residential construction. To avoid such a narrow perspective, a generalized system of construction is developed that applies to all constructed products. Often this starts with a breakdown of the overall systems model of construction into subsystems. For example, one may conclude that all constructed products have a "foundation subsystem" which is the basic support for the constructed product or, in other words, the interface or connection between the constructed product and the earth. The foundation subsystem applies to all

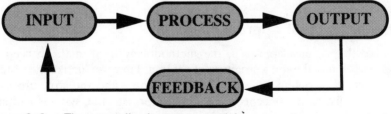

Figure 8–3:   The generalized systems model.

constructed products, including roadways, causeways, buildings, bridges, towers, and so on. Thus, content-wise, the nature of the earthen materials upon which the structure is to bear and their properties, as well as the practices to distribute loads to prevent settling, would be some of the universal topics studied using the systems approach that are applicable to all constructed products.

The methods used in the systems approach potentially cover the entire range of possibilities, depending upon the content that results from an analysis of construction as a system. The important methodological criteria is that the students apply systems thinking to their analysis of construction designs, problems, and learning activities. Moreover, the methods must relate directly to the universal nature of the ideals of the systems approach. In other words, activities must result in learning that applies not just to residential construction, but to all forms of construction. The extent to which the ideals of the systems approach are realized varies greatly among construction curricula and instructional materials.

The systems approach has two major advantages. First it gives the students a way of thinking about the human-built world in an organized and logical manner. Thus, when confronted with a new and unfamiliar situation (constructed product), the student can analyze it in terms of the universals of systems theory and thus understand it. Second, it gives the teacher a tool to analyze construction content and use methodology that is consistent with that content. This moves the content and resultant methodology away from a delimited focus upon residential construction and selected trades.

The biggest disadvantage to the systems approach is realizing the advantages. Doing so is a methodological problem rather than one of content. The systems approach can provide the teacher with a universal approach to construction content. But much work remains in hands-on, engaging activities that bring the content to life for the student. A lot must be done to meet this challenge. Application of the systems approach must go far beyond having the students memorize the elements of the systems model. The systems approach is principally a content approach.

The Conceptual Approach

Like the systems approach, the methodology used in the conceptual approach is based upon a structure of content. However, unlike the systems approach, the content is usually more definitive and the structure is an organized listing of concepts (rather than, for example, systems and sub-systems). In addition, a conceptual structure (as well as a systems structure) is not simply a topical outline of the content.

One of the classic examples of the conceptual approach applied to technology education was the Industrial Arts Curriculum Project (IACP). The forerunner of the IACP *The World of Construction* program was the development of a *Rationale and Structure for Industrial Arts Subject Matter* (Towers, Lux, & Ray, 1965). All of the concepts in the IACP structure were "verbal nouns or gerunds, which connote 'action' or 'doing'" (p. 169). Thus, all of the elements of the structure ended in "ing." For example, preparing the site, excavating, and installing utilities are concepts found in the IACP structure. The reader of the content structure has a clear idea of what kind of doing is intended.

Assuming that the content structure is inclusive of all significant and appropriate construction concepts, then the methodological task is to develop activities for students that teach the concepts included in the structure. This gives the teacher and other curriculum developers a clear and consistent point of reference. And, if the structure is truly based on concepts, then the inclusion of meaningless facts and rote memorization is reduced. However, developing meaningful activities that are totally inclusive of the conceptual framework can be a very difficult undertaking. As a result, the content structure may be top-heavy in cognition to the exclusion of the other domains of knowledge and emphasize the lecture as the principal instructional method. Or the concepts may be delimited to those for which a correlated activity can be developed.

Another potential pitfall of the conceptual approach is that the product that results from the activity can become an end in itself. For example, a major activity in *The World of Construction* program was the Construction Module. This activity was carefully designed to provide a wide variety of construction experiences. Management (and personnel) practices were essential parts of what the students learned by actually engaging in the major activities that construction managers did (planning, organizing, and controlling). Role playing was used extensively. But, as with many curricular innovations, entropy set in Management concepts, an essential part of the program, were eliminated. Textbook reading assignments were discarded and other key experiences were cut. In some schools, the program degenerated to the "here are the plans - build it" approach that the field has tried to evade for decades.

The problem of confusing ends and means is certainly not a new dilemma in technology education nor is it exclusive to the conceptual approach. In the case of *The World of Construction* and similar programs, there are a number of possible explanations for the confusion. Cost is one - the materials required to build the modules were expensive. Second, the program takes considerable preparation time and effort on the part of the teacher. But most important may be the lack of a clear philosophical base upon which to make educational decisions. Because of this, the adaptations made by the teacher may have been done with good intention and without the realization that, in many cases, the changes completely bifurcated the program and its educational integrity. It is an interesting example of how modifications in method result in a destruction of content. For more information on the utilization of *The World of Construction*, see LaPorte (1980).

The Simulation Approach

In simple terms, the simulation approach attempts to present a model of artificial reality to the student. The student plays the role of a person within this artificial reality. Many games place the game players in a role which simulates reality. Monopoly for example, places the game player in the role of an investor with the goal of maximizing net worth. Likewise, the Big Builder game from *The World of Construction* can put the student in the role of a real-estate developer. Simulations have realized tremendous success and learning effectiveness. Very sophisticated simulations are used to model business problems.

The simulation approach need not be limited to published simulation products. In fact, teachers often use the approach without even realizing it. For example, teams of students are often given the responsibility for carrying out a particular construction project under the direction of a team leader. The team leader assumes the role of the contractor and the other students play the roles of subcontractors or skilled craftspersons. Another example is the simulation of labor negotiations in which some of the students represent management and others represent labor. A student may play the role of arbitrator. Even organizing lab cleanup in which one student is the foreperson and other students are assigned specified cleanup duties is, in effect, a simulation.

The advent of the computer dramatically changed simulations and simulation games. A simulation game called Lunar Lander appeared on mainframe-computer systems in the late 1960's. In this simulation, the game player inputs various rocket-thrust values in an attempt to make a safe landing on the moon under the influence of increasing lunar gravity. Lemonade was a simulation that appeared on early microcomputers. Based on economic decision making, the goal of Lemonade was to try to match the supply of lemonade with demand and maximize profit. Weather was a random variable. These early computer simulations were, for the most part, text-based.

Aircraft-pilot-training simulators, controlled by mainframe computers, spurred the development of computer-graphic simulation. Shortly after the introduction of microcomputers, computer-graphic simulation games were developed. Though crude at first, these games developed into very sophisticated simulations of the real world (or a world contrived by the software developers). Among the most realistic computer-graphic simulations of technology is Flight Simulator (Microsoft and Sub Logic corporations).

Sophisticated simulations are available to teach both physics and chemistry. They are designed to teach classical experiments in these sciences. The simulations allow for a great deal of interaction between the student and the computer. In the chemistry simulations, for example, students have control over such variables as the amount and type of chemicals to be used, laboratory apparatus setup, and heating. Sensory feedback is provided to the student, both visually and audibly.

Unfortunately, very little software, let alone computer graphic-simulation software, has been produced expressly for technology education. What little does exist is, for the most part, drill and practice nomenclature. However, there are some software packages, albeit designed for other subjects and purposes, that can be used in construction. One is SimCity (Maxis Software). In SimCity, the simulation player assumes the role of a mayor of a community. Using a tax base for resources, the goal is to develop a thriving community in which the citizens (Sims) are happy and prosperous. Decisions must be made about such things as the economic base (commercial or industrial), housing, transportation, police and fire protection, and the tax rate. Though SimCity is targeted toward social science teachers and their students, the program can be readily adopted to fit within a unit on city and regional planning in a construction course. Teams of students can compete with one another to see which team can develop the most prosperous city. The program can be an excellent lead-in to the need for zoning ordinances and building codes. The students can also see the relationship between economics, government, and construction. SimCity has won several educational software awards.

Another simulation that can be used in construction is called SpaceMax (Final Frontier Software). The goal of SpaceMax is to build a space station. The simulation player must decide what tools and components, as well as the human resources (and the food and oxygen they require), must be taken on each trip into space. Economics play an important part in the simulation and budget constraints are a reality. As with SimCity, students can be organized into teams to meet the challenge of building the space station and the teams can compete with each other to see which teams can meet the object most economically.

SpaceMax is a rather sophisticated simulation. It requires that the students know the details of how the space station is constructed and how each component fits with one another. In fact, each component is depicted graphically and it must be linked up properly with the other components with which it fits. Though this seems to appeal to students and it gives them the sense of actually building something, it requires a considerable amount of time for them to learn exactly how the space station goes together. A large-scale plan, which, unfortunately, is not supplied with the software package, would make this task much easier.

Computer-graphic simulations offer much potential in helping construction programs move away from the emphasis on residential construction. What is missing are software developers who are willing to embrace technology education and develop pertinent software for the field. For example, the concept used in SpaceMax could be applied to a variety of constructed products, from houses and commercial structures, to bridges and skyscrapers. How to get software developers to support technology education remains a challenge.

"Virtual reality" appears to be the next step in simulation development. Already, through the use of special goggles, participants can experience a simulated reality in which they can walk and look anywhere they please with both visual and auditory feedback. With the further development of ideas like force-feedback gloves (see Brand, 1988), tactile feedback will also be a part of the virtual-reality experience. Virtual reality of sorts is already available in such architectural computer-aided design (CAD) packages such as DataCad (Cadkey, Inc.). After the development of a set of plans for a building, DataCad allows a person to "walk through" a visual representation of the building on the computer monitor. Thus students can experience a limited virtual reality experience right now as part of their CAD experience. Teachers could also develop various constructed products and then have students explore their details through the "walk-through" concept (or commercial developers could develop such software).

It is quite plausible that, once virtual reality is developed further, students will be able to "build" a variety of structures. They could actually learn about various materials, including their textures (through the force-feedback glove mentioned earlier). They could then assemble the materials, "virtually building" whatever their imaginations can come up with. Then, their designs could be tested and evaluated and then modified to improve them through several iterations.

Non-computer-based simulations, when appropriately applied, have a fairly substantial body of research evidence to back up their effectiveness. Computer-graphic simulations have some research evidence to support their effectiveness as well (especially in non-public school educational settings like pilot training). However, very little research has been done regarding computer-graphic simulations in technology education. Of course, this is of little surprise since the simulations must be developed before research can be conducted. As computer-graphic simulations are developed, it will certainly lead to consideration of what "hands-on" really means, a question that is just beginning to be pondered within the field and one which requires a rigorous research agenda for its answer.

The Model Building Approach

Model building has been used extensively in technology education. During W.W.II, industrial arts students participated in the war effort by building models of planes that were used to train military personnel in air combat tactics (Groneman, 1944). Maley, in the *Maryland Plan* (1973), utilized model building extensively. The IACP World of Construction program used model building in the ever-popular Dream House activity. An adaptation of the Dream House is still very popular in state, regional, and national technology education exhibitions and contests.

Models showing the structural details of how a house is constructed have been built by students for decades. Architects and engineers still rely upon models to show, three-dimensionally, the details of a variety of constructed products despite the availability of dynamic, photo-realistic displays available through computer technology (though the reliance upon real models is gradually diminishing). Museums rely upon models to a large extent. Models also have a strong attraction to people of all ages. Judging from the continued attractiveness of doll houses, model trains, and model kits, models appeal to young people and do not appear to be gender-specific.

In construction, model building seems to be used in two principal ways. First, models are used as the vehicle for design expression (e.g., the Dream House), which might be called aesthetic models. Like the architectural model, the aesthetic model emphasizes appearance and design. The second type of model is the structural-detail model. An example of this type is a model of a house in which all the framing members, utility details, and foundation details are shown.

The use of models in technology education seems to have increased as programs move to the broader scope of technology education. In construction, this broader scope has been a move away from the almost exclusive emphasis on residential construction. Concomitantly, in a search for activities that fit the breadth, models have become an alternative (the sole alternative?). For example, if the students are studying different types of skyscrapers, it is difficult or sometimes impossible to find meaningful activities other than model building.

In addition to a broader curriculum, another hallmark of technology education has been an identifiable movement to devalue the teaching of skill in the use of tools and machines (e.g., Yu, 1991). This skill was formerly the most highly valued objective of industrial arts (Dugger, 1980; Schmitt &

Pelley, 1966). With the de-emphasis on tool skills, model making has become a viable laboratory activity.

There are some disadvantages to the model-building approach. First, it can easily be overdone. There is a rather natural interest among students in model making, as mentioned earlier. However, this interest can be easily destroyed through repetitive model-making experiences. Second, younger students may not have the psychomotor skills to produce meaningful models. They may find the experience frustrating and boring. Building a model from scratch based on research that the student has done is quite different from assembling a model kit in which specific directions are provided. Third, the real learning about the subject represented by the model may not be efficient relative to the time required. Other methods of teaching about the subject of the model, such as video tapes, written materials, or other media, may be more efficient and even more effective than building the model. Students may become so engrossed in how to make the model that they lose sight of the learning that the model was intended to facilitate.

The Design-Brief Approach

The design-brief approach is an adaptation, for the most part, from work done by the British in their Crafts, Design, and Technology (CDT) program (now referred to as the Design and Technology Program). The student is given a synopsis of a design problem in a written description called a design brief. The brief generally consists of standard sections. The situation sets the context for the design problem. It is the link between the problem in which the student will be engaged and the real world. The problem statement is a concise description of the problem to be solved. The evaluation section describes the criteria for evaluation and the conditions under which evaluation will occur.

The design-brief approach relies heavily upon the portfolio-assessment technique. The portfolio is essentially a diary of how the student approached the solution of the problem and the thought processes used. The portfolio relies heavily upon graphics to depict the evolution of the problem solution. The quality of the graphics in examples done by British students is generally very good whereas the graphics in the Americanized examples tend to be rather basic and relatively crude. This is no doubt due to the much greater emphasis placed upon teaching graphic-rendering skills in the British system. The portfolio is designed to assure that evaluation focuses upon the process of solving the problem rather than upon just the end product.

The design-brief approach emphasizes the application of a problem-solving model in which technological problem solving is done by following a sequence of processes or steps. The original CDT model consisted of the following steps (Finney & Fowler, 1986):

1. What do we intend to design and make? (problem statement).
2. Research and ideas.
3. Development of chosen ideas.
4. Working drawing and planning procedure.
5. Making.
6. Evaluating.

A number of problem-solving models have appeared in the literature both in this country and abroad. For the most part, they are variations on the same theme. But application of the problem-solving model and use of the design-brief approach have become for some the *sine qua non* of the transition to technology education. It is interesting to note the similarity between the CDT model presented above and the steps that Lindbeck proposed in his design textbook over 30 years ago (1963, p. 106):

1. Statement of the problem.
2. Analysis of the research.
3. Possible solutions.
4. Experimentation.
5. Final solution.

Figure 8-4 is an example of a construction design brief:

With the design-brief approach, the student has a clear conception of just what the problem is. Through application of the technological problem-solving model, the student learns a logical approach to the solution of the problem. The student clearly sees the need for library research, consultation, and experimentation. In addition, the approach parallels what actually happens in trying new designs and approaches in the construction industry. If portfolio evaluation is used, there is some assurance that the students actually applied the problem-solving model. The competition that occurs in the evaluation of problem solutions can increase the motivation of the students and allows those who are not successful in other academic areas to achieve success. It promotes real thinking and analysis.

The example given is not inclusive of all construction-related design briefs; it seems to be typical of many that are being used in schools at the present time. Similar briefs include bridge building, cantilever beams, and

Design Brief
Water Tank Supportive Structure

Situation

As an engineer, you have been asked to design a supporting structure for a water tank. The tank will serve as a reservoir for a new community. The contract for the reservoir will be awarded to the engineer that develops the structure that supports the greatest amount of water with the construction materials given.

Problem

Given 40 toothpicks and white glue, develop a water tank supporting structure that will elevate the bottom of the tank 3 inches above a table top. The bottom of the tank is 2.5 inches in diameter (a discarded food can).

Evaluation

The can will be placed on the supporting structure and weight will be added until the structure breaks. The strongest structure will be the best design.

Figure 8–4: An example of a construction design brief.

other structure problems. There are several problems with many of these structure-related design briefs which can be illustrated with the water-tank example. First, the toothpicks which are used as the primary construction material may be too small for younger (i.e., middle school) students. Second, many potential variables can explain why a structure failed or succeeded other than the design of the structure itself. The time that elapses between when the glue was applied and when the members were joined affects the ultimate strength. The strength varies dramatically from one toothpick to another, much more so than do real structural materials. Moreover, the strength of scaled-down wood materials can not be generalized to their full-size equivalents since the wood fibers remain the same size. The net result is that students may often make erroneous conclusions about the structural-design concepts that are the intent of the activity. The net result is that they may develop misconceptions about what was to be learned comparable to those about which the science community has been concerned in recent years.

But there are also more significant problems that appear to be occurring with the design-brief approach, no doubt to the chagrin of those promoting the approach. Often students are expected to solve technological problems with the knowledge with which they entered the technology course. In other words, they have no knowledge base upon which to solve the problem. Learning could perhaps occur if the students could try an idea, test it, redesign it, and test it again. However, the testing is often a one-shot experience, leaving the student with more questions than answers and minimal gain in knowledge. The iterations which could lead to knowledge are eliminated. But there are practical reasons why this iteration does not occur; it takes a lot of instructional time and students may become quite bored with doing the problem repeatedly.

In days gone by, teachers used to share plans for student projects at professional meetings and elsewhere. These projects plans, sequenced by the teacher, became the curriculum. There is evidence that the same phenomena is occurring today (as mentioned earlier) with sharing design briefs. The students simply do one design brief and then the next. There is no articulated program, no real goals, no real objectives. Even the learning that occurs through solving the problem stated in the design brief may be questionable. Such a program is simply not defensible.

Finally, the scaled-down nature of the problems typically addressed in the design-brief approach calls into question the need for the technology laboratory since many popular design briefs could be done in a regular classroom. With the typical press for physical space in most schools today, if the equipment and space within the technology laboratory is not used effectively, it will no doubt be taken away.

The design-brief approach has been discussed in greater detail than the other approaches, primarily because of its pervasiveness in technology programs today. The approach has many attributes worthy of consideration in a construction program. However, caution must be used so that it is implemented in a way that it results in valuable learning experiences for students and is part of a well-defined and articulated curriculum.

CONCLUSION

If the reader has the impression that the writer believes that an operational set of approaches to teaching construction in technology education do not yet exist, that impression is valid. As mentioned earlier, the development of content structures for construction is leading the development of appropriate instructional approaches. The content found in curriculum guides and in textbooks is broad enough to encompass the totality of construction. However, for the most part, the instructional approaches used still either emphasize principally residential construction or the learning gained from the approaches is of questionable value and relevance.

It is interesting to note how little literature has been written about construction in technology education in the past decade and the dearth of presentations at professional conferences. Most of what has been written about construction education focuses upon vocational-trade preparation, especially as it relates to special populations. One could surmise that either the profession believes that construction has been developed to an ideal level already, there is a declining interest in construction, or few have been able to meet the challenge of developing new approaches. Hopefully, the latter is the valid supposition.

The fact that many of the selected approaches to teaching construction emphasize primarily residential construction has been mentioned repeatedly. It seemed logical to do so since this has been a criticism for a number of years. However, it is interesting to reflect upon the industrial arts years for a moment. In those earlier years, goals relating to leisure time and consumer knowledge were essential components and residential construction was a natural emphasis. But one would have a difficult time finding but a few references to these goals in our literature during the past decade.

No doubt about it, the leisure time aspects of earlier programs were emphasized far too much in the past, either due to philosophical leanings or due to ignorance about how to move away from such an emphasis. It is a welcome trend, indeed, to see some change. Yet, there seems to be a growing interest among the general population in do-it-yourself work to save money, for the joy of it, or both. This trend is evidenced by the growing number of periodicals and books devoted to the subject that appear in bookstores. Should technology education programs address worthy leisure time objectives in a construction program? Is this a defensible undertaking? If so, how can it be approached so that it does not become a principal focus once again? Other teachers aspire to develop leisure time interests that relate to their disciplines. What language teacher would not be pleased that they influenced students to read for pleasure? What physical education teacher would not be proud that they motivated students to have a lifelong commitment to staying fit? Or what art teacher would not relish the fact they developed an appreciation in art?

The American dream is still typified by owning one's own home. For most, such ownership will be the biggest financial decision that a person will ever make and, if not made on the basis of knowledge, could lead to very serious personal hardship. Does technology education still have a responsibility to educate students to be wise consumers of constructed products? If so, what is that responsibility and what approach should be used to meet it?

Interestingly, it is with some trepidation that the writer even mentions leisure time and consumer aspects, for fear of being chastised as a traditionalist. Yet, it seems quite logical that these two areas are no less important than when they were first subscribed to years ago. Nor, does it seem likely that the same rationale for their inclusion in the past would not apply today.

The residence in which our students live has far greater immediate meaning and relevance than do bridges, skyscrapers, and causeways. We do need to assure that they are literate about the systems within that residence before studying other constructed products. This may explain the apparent continued emphasis on residential construction and offers some rationale for it. Once this literacy has been developed, then instruction can proceed to the larger constructed world. A very logical approach, indeed, but it ignores the fact (as we often tend to do) that technology education is, by and large, an elective course in the schools. There is no hope, at least presently, of providing an articulated program. By choice or not, construction is an open-entry, open-exit program. Thus, we must provide the breadth of comprehensiveness and the depth to avoid superficiality in a one-shot approach. This, indeed, is a challenge and an impediment in realizing our ideals.

How to realize our ideals in the most effective manner begs for research. So does the validity of the various approaches presented. What works and what doesn't? In what situations do certain approaches work better than others? Is it possible to emphasize residential construction, and use it as a meaningful referent to teach broader concepts? If the literature is replete in construction, the research is doubly so.

Generalizing from the work of Yu (1991) (mentioned earlier), skill in the use of tools and machines is apparently regarded relatively low in importance by professionals in technology education today. If this is the case, then how does one know when hands-on activities are appropriate (recognizing that they take more time) or when some other approach such as using video tapes, computer software, or other media is more effective? Even the whole idea of hands-on learning suffers from a lack of research to substantiate it. Again, research is needed.

So, technology education has a research dilemma. And the dilemma is difficult to solve. Enrollment in teacher education programs is at an all-time low (see Householder, 1992). This means that there are fewer teacher educators in general and those who remain often find themselves principally involved with programs and activities outside of the field. And these very teacher educators are the source of research-based knowledge. Grass roots, action research by practicing teachers may offer an alternative solution.

The book on approaches to teaching construction is certainly not closed. Much work needs to be done in developing new approaches to teaching construction, determining their effectiveness, and disseminating them to the profession. Construction, in fact, may be the disadvantaged child of technology education. The parents assumed that everything was in order and gave the child little attention. In reality, the child has some deep-seated problems which can be eliminated with some attention.

REFERENCES

American Association for the Advancement of Science. (1989). *Science for all Americans: A Project 2061 report on literacy goals in science, mathematics, and technology.* Washington, DC: Author.

Bensen, M.J. (1980). Selecting content for technology education. *Proceedings of Technology Symposium 80* (pp. 12–14, 24). Charleston, IL: Eastern Illinois State University.

Bernhardt, L.J., & McHaney, L.J. (1992). *Space simulation.* Albany, NY: Delmar.

Bertalanffy, L. von. (1962). *Modern theories of development.* New York: Harper Torchbooks.

Brand, S. (1988). *The media lab.* New York: Penguin Books.

Brandt, R.S. (Ed.). (1988). *Content of the curriculum.* Reston, VA: Association for Supervision and Curriculum Development.

Claxton, C., & Murrell, P. (1984). Developmental theory as a guide for maintaining the vitality of college faculty. In *Teaching and Aging.* San Francisco: Jossey-Bass.

Dugger, W.E. (1980). *Report of survey data - Standards for industrial arts programs project.* Blacksburg, VA: Virginia Polytechnic Institute and State University.

Finney, M., & Fowler, P. (1986). *Craft, design, and technology foundation course.* London: Collins Educational.

Groneman, C.H. (1944). Model aircraft program proves successful. *Industrial Arts and Vocational Education, 33*(2), 117–118.

Guild, P.B., & Garger, S. (1985). *Marching to different drummers.* Alexandria, VA: Association for Supervision and Curriculum Development.

Henak, R.M. (November, 1992) Effective teaching: Addressing learning styles. *The Technology Teacher, 52*(2), 23–28.

Henak, R. (1993). *Exploring construction.* South Holland, IL: Goodheart Willcox.

Householder, D.L. (1992). *Assessing teacher education programs.* Paper presented at the 79th Mississippi Valley Teacher Education Conference, Chicago, IL.

Kolb, D.A. (1984). *Experiential learning: Experience as the source of learning and development.* New York: Prentice-Hall.

LaPorte, J.E. (1980). The degree of utilization of Industrial Arts Curriculum Project materials relative to their perceived attributes, teacher characteristics, and teacher concerns. Unpublished dissertation, The Ohio State University.

LaPorte, J.E., & Sanders, M.E. (1992). The T/S/M Integration Project: Integrating technology, science, and mathematics in the middle school, *The Technology Teacher* (accepted for publication).

Lindbeck, J.R. (1963). *Designing today's manufactured products.* Bloomington, IL: McKnight.

Lux, D.G. (1979). Trade and job analysis: The scourge of industrial arts. *School Shop, 38*(7), 2.

Lux, D.G. (1977). From heritage to horizons. *Journal of Epsilon Pi Tau, 3*(1), 7–12.

Maley, D. (1973). *The Maryland plan.* New York: Bruce.

McCade, J. (Ed.). (1992). *Proceedings of the 1992 Technology Symposium.* Millersville, PA: Millersville University.

National Council of Teachers of Mathematics. (1991). *Professional standards for teaching mathematics.* Reston, VA: Author.

Pucel, D.J. (1992). *Technology education: A critical literacy requirement for all students.* Paper presented to the 79th Mississippi Valley Industrial Teacher Education Conference, Chicago, IL.

Riner, P.S. (1990). Authentic education: Resolving the process-content dichotomy in curriculum debates. *Journal of Humanistic Education and Development, 29*(1), 61–68.

Schmitt, M.L., & Pelley, A.L. (1966). *Industrial arts education: A survey of programs, teachers, students, and curriculum.* (Circular No. 791, OE

The Secretary's Commission on Achieving Necessary Skills. (1991). *What work requires of schools - A SCANS report for America 2000.* Washington, DC: U.S. Department of Labor.

33038). Washington, DC: U.S. Department of Health, Education, and Welfare.

Snyder, J.F., & Hales, J.A. (1981). *Jackson's Mill industrial arts curriculum theory.* Charleston, WV: West Virginia Department of Education.

Towers, E., Lux, D.G., & Ray, W.E. (1965). *A rationale and structure for industrial arts subject matter.* Columbus, OH: The Ohio State University Educational Foundation.

Witkin, H.A. (1976). Cognitive style in academic performances and in teacher-student relations. In S. Messick (Ed.), *Individuality in learning.* (pp. 27–38), San Francisco: Jossey-Bass.

Yu, K.C. (1991). A comparison of program goals emphasized in technology education among selected groups of professionals in the state of Virginia. Unpublished doctoral dissertation, Virginia Polytechnic Institute and State University.

CHAPTER 9

Facilities For Construction In Technology Education

Douglas L. Polette, Ed.D.
Department of Technology Education
Montana State University, Bozeman, MT

The world is going through major transformations in all aspects of technology. The major changes that are taking place in the teaching process and in classrooms and laboratories make it imperative that high-quality facilities are designed and built. Recommended learning strategies engage the student in their own education through a variety of individual and group activities. They are specifically designed to provide the student with an understanding of the major concepts pertaining to the topic under study. The technology education laboratory is an example of a modern learning environment where the student can be a creative problem solver who uses a variety of resources in a stimulating environment.

Construction technology requires a well-designed laboratory in order to carry out the physical activities needed to help the student move from concrete experiences to abstract concepts. This approach allows the student to assimilate the knowledge presented into practical, usable forms. Although each construction laboratory design will vary with the level of instruction, the major concepts of construction laboratory planning remain consistent throughout all levels of education. Today, more than ever, the facility must be designed with the future in mind. It is not easy to look into the future; but there are principles that assure the construction facility meets the educational needs of today and tomorrow. In this chapter, these principles will be addressed and provide the reader with a useful procedure for designing high-quality facilities.

PHILOSOPHICAL FOUNDATIONS

The need for a philosophical support for construction in technology education activities is well documented (Hales & Snyder, 1981; Helsel &

Jones, 1986; Savage & Sterry 1990). Regardless of the curriculum organizers used, construction activities play a major role in curriculum design. Throughout history, construction has played an important part in the development of all societies and will continue to play a major role in the future. Citizens cannot participate in a democratic society without having some understanding of construction systems. It is imperative that students receive a better understanding of construction. Activities must deal with more than just stick-built residential construction. Students must develop an understanding of how roads, bridges, dams, docks, utility systems, commercial, and industrial construction impact their lives. Broadening the construction base beyond traditional activities will require a laboratory which is different from those designed for industrial arts construction programs.

CHARACTERISTICS OF THE CONSTRUCTION FACILITY

Curriculum Structure For Construction

Instruction in construction topics should occur in a facility that is specifically designed for the purpose of addressing construction in technology education. The physical arrangement and size, shape, tools, and materials used will depend upon the educational level of the program. However, the first and most important prerequisite is that the facility design follows the program philosophy and curriculum design for the construction in technology education program. Without a solid philosophical basis, and a well-thought-out curriculum design, the laboratory will not meet the needs of the program or the learner. The construction program must be based on a taxonomy that will provide for both current and future programs. The taxonomy in Figure 9–1 can help in developing the construction in technology education curriculum. The course content should address the basic concepts of the total field of construction as described in earlier chapters. A comprehensive construction course would draw from the entire taxonomy of construction and the laboratory must be designed to complement the curriculum design.

The Planning Process

Once the curriculum is designed, completed, and assessed to assure it meets the goals of the technology education program, it is time to assess the physical environment where instruction will occur. These spaces include a

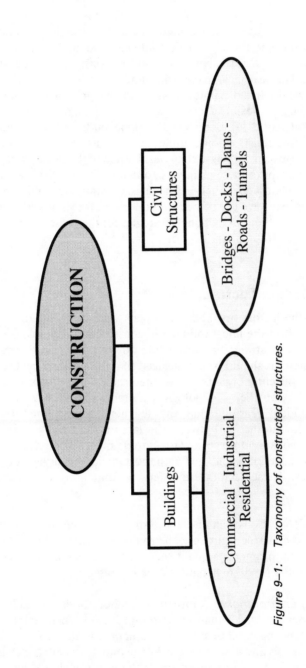

Figure 9–1: Taxonomy of constructed structures.

classroom, laboratory, and auxiliary spaces. The curriculum design will help to define the activities implemented in the construction laboratory, and therefore determine the equipment and space needed. Regardless of the program level, the physical arrangement of the construction laboratory must be designed with learner outcomes in mind.

Because we live in a dynamic era where change may be the only constant, the facility must be flexible. Certainly, a variety of changes are sure to come (Daiber & LaClair, 1986). Because educational budgets seldom allow for the latest up-to-date construction equipment, we must be able to simulate new developments in a cost-effective manner. Design of the facility will, in a large part, have to adapt to these changes in a manner that will allow students to experience the latest construction practices, yet stay within a limited budget. This requires extensive use of video tapes, computer simulation, interactive instruction, and other instructional strategies for the presentation of current information. A careful integration of all available resources will result in the best value.

Overall Design Factors

Some of the factors that come into play when designing or redesigning facilities to reflect the latest thinking in construction technology education are mundane and straightforward, while others take years to develop and fully implement. It should be emphasized that the physical facility is simply a means to efficiently organize tools, materials, and processes so students can better develop their technological literacy. The facility can and does influence student learning and the teacher's instructional techniques. Therefore, the interaction between the laboratory, activities, courses, and perceptions the laboratory convey to learners must all be integrated. Three overall design factors stand out as important in the design of construction facilities. These factors are safety, a stimulating environment, and flexibility.

Facility Safety. Construction laboratories must be safe for both the student and the instructor. A safe environment is central to establishing an atmosphere that promotes a quality educational program. If it is not a safe environment, the program should not be operated.

Stimulating Environment. Construction laboratories should be a stimulating environment that enhances learning. Include such simple things as wall charts, labeling of all pieces of equipment, and the proper storage and placement of tools and machines, etc. When one walks into the construction laboratory, the immediate impression of the purpose should be present in the observer's mind. Does the facility communicate to the person? Is there curiosity about the construction tools, processes, and techniques? The facility must communicate, entice, and invite students to explore this exciting area of technology.

Laboratory Flexibility. The importance of designing flexibility into technology laboratories has been well investigated by previous authors such as (Cummings, Jensen, & Todd, 1987; Braybrooke, 1986; Daiber & LaClair, 1986; and Brown, 1990). These authors stated that flexibility must be a major consideration in the design of new laboratories. The potential for change in the specific laboratory arrangements of equipment and instructional activities necessitates that the facility layout can be changed with a minimum of time, effort, and costs. It is important to select equipment and furniture that is portable which in turn influences electrical, air, and exhaust systems.

OVERVIEW OF THE CONSTRUCTION TECHNOLOGY FACILITIES BY LEVELS

Just as the curriculum will vary at each level of the school system, so must the facilities be adjusted to best meet and support the instructional goals of the program. This section addresses how educational levels influence the construction environment.

The Elementary Program

K-5 students need to develop an awareness of the importance of construction technology in everyday life. They need to begin to develop an understanding of the tools, materials, processes, and outcomes of the construction enterprise. The best way for them to gain this understanding is through hands-on activities that simulate the construction technology field in an interdisciplinary setting. Construction activities should be introduced at the elementary level to enhance the student's technological literacy. Construction activities provide a natural setting for the student to use mathematics, science, language arts, and social skills together in activities which enhance their understanding of integrated disciplines.

At the elementary level, children should become aware of the vast size of the construction industry and that our built environment is totally dependent on the construction industry. They should learn that, throughout history, humankind has relied on construction activities to provide shelter, build transportation systems, and aid in the development of society. They should be able to identify the resources, the major construction systems, and the impact of construction technology on society and the environment.

The elementary construction laboratory does not have to be a separate space; it can be part of the classroom. The size and capabilities of young students need to be kept in mind when materials, tools, and equipment are acquired. Often materials such as cardboard, soft woods, plastics, and light metal are the materials of choice. Some or all of these materials may need to be pre-processed by the teacher before they can be safely, conveniently, and efficiently used by elementary students. An elementary teacher should work closely with the technology teacher at the middle school or high school to assist in curriculum design and the pre-processing of the materials. It is desirable to have portable tool carts, work surfaces and basic equipment that can be shared by all elementary teachers in a school building. This equipment does not have to be extensive but should be matched with the age and physical strength of the elementary student.

The Middle School Program

Construction technology experiences at the middle school level need to provide for gradual transition from the typical self-contained elementary classroom to the highly departmentalized program at the high school level. In writing about the structure of the middle school, Bame (1986) listed five characteristics of the middle school program: "The curriculum should be: exploratory, understanding, broad and fundamental, interdisciplinary, and vertically integrated" (p. 73). Construction activities at the middle school should address these characteristics to determine the type of laboratory needed.

A middle school laboratory should include a variety of equipment, tools, and materials typical of technology education. Since exploration is the primary goal of the middle school technology education program, a separate construction laboratory for this level is not warranted. Rather, the technology facility must be flexible enough to promote integration of the various technology clusters. Also the integration of other disciplines will enhance the students' overall understanding of our technological world. Current emphasis on problem solving in all curricula necessitates that the laboratory be set up with a variety of activities.

Bench-top equipment for a middle school program can be adequate if it is of high quality. At this level, students should begin to take on the responsibility of operating certain power equipment in a safe and efficient manner. Consequently, the equipment, both computer and material processing, needs to be properly selected so a beginning student can operate it with a minimum of risk to themselves, others, and to the equipment. A multitude of materials should be available to the student so they can begin to assess strengths, weaknesses, and other properties of materials used in the construction industry.

The Senior High School Program

A high school laboratory should be designed to support the approved curriculum for students of this age group. Depending on the size of program and curriculum design, the laboratory may be a separate facility or combined with the manufacturing cluster as part of a production laboratory. Whatever the decision, the construction portion of the program needs to be set up so the high school student is provided with a comprehensive view of construction technology. The materials, tools, equipment, and processes need to be broad enough to provide comprehensive experiences in the construction industry and extend beyond residential stick-built construction typical of the past.

Students should have opportunities to develop career awareness in the total field of construction technology. This can be accomplished by including experiences in the construction of roads, bridges, dams, docks, pipelines, and a variety of buildings. The laboratory required to conduct these activities needs some traditional equipment to process wood, metal, plastic, and ceramics. It also requires an investment in computers and software that allows students to explore design and management activities. Activities such as the production of asphalt used in the construction of roadways or airport runways may not be practical to replicate. With instructional media and field trips, students are able to develop an understanding of the important aspects of construction too expensive, too far away, or too hazardous to experience personally. Standard equipment such as separating, combining, forming, and conditioning equipment needs to be supplemented with a variety of testing and measuring equipment. Students are then able to test and evaluate a variety of materials and construction techniques to determine an appropriate solution for a given problem.

The program must be based on concepts of construction rather than the development of specific trade skills; and, equipment selection must contribute to development of the concepts. The purchase of multiple pieces of equipment such as radial arm saws and other costly pieces of stationary equipment should be discouraged. The exception would be computers because the availability of a large number of software programs makes each computer a different learning station and computers can be infused throughout the curriculum.

There should be a clean and quiet space as well as spaces in which materials are processed where noise, smoke, chips, and sawdust are produced. Computers and related equipment used to enhance the construction experience should be placed in an isolated portion of the laboratory away from the dirt and dust.

The Teacher Education Program

The teacher education laboratory should reflect the latest in trends of construction technology and the philosophy of technology education. Various laboratories need to be grouped together to enable students to experience the integration of clusters and the professional teacher education sequence into an efficient, cohesive program. In addition to the construction laboratory, there should be a variety of support spaces such as a seminar room, resource center, standard classroom, and computer laboratory. Keep the facility flexible; it will allow for maximum utilization of the space and for future growth.

Whether a construction laboratory may or may not be a stand-alone facility depends on the size of the teacher eduction program and its link with other closely allied programs. If the construction laboratory cannot demonstrate student usage for the majority of the day, it may not be possible to generate the administrative support to have a dedicated construction laboratory. In this instance, laboratory space may have to be shared with an allied program cluster such as manufacturing.

Teacher education programs should provide students with a construction laboratory that allows them to engage in a comprehensive construction experience and provides an opportunity to practice teaching strategies. Construction laboratories need to be adaptable and responsive to the changing demands of both teacher education and the construction industry. In construction laboratories, future teachers develop technical skills and understandings of the construction industry. The facility should be equipped with the latest construction tools and equipment including testing and measuring equipment that provides students with experiences that are current with today's construction practices. A variety of computer hardware and software should be available to give students opportunities to become technologically literate with the latest construction practices and standards before they enter the teaching profession.

The Construction Laboratory And Public Relations

Any laboratory can be a show place for the school. School administrators and teachers should be able to show visitors how the school program is providing quality education. A standard classroom with rows of desks, a chalkboard, and typical resources on the walls does not impress the visitor. Visitors want to see dynamic educational settings where students are actively engaged in learning experiences. Well-designed, efficiently organized, and neatly kept facilities can leave a positive impression with administrators, visitors, and the general public. Much of the public is confused because many educators have just changed the name over the door and little, if any, real change has taken place in the curriculum or the laboratory. The design and layout of the facility as well as the learning activities provide the impression of the program. This image communicates if the program is on the forefront of technology education and if the students are getting a quality education. The visual appearance of the laboratory can create interest and curiosity or close the visitor's mind.

Subject Area Integration

Subject area integration is another common thread that must run through construction in technology education programs. No single discipline or part of a discipline can justify its existence on content alone. There must be integration within the discipline and with other disciplines that are under study by the student. Construction technology relies on mathematical and scientific principles, social interaction, communications skills, aesthetics, and other disciplines. The curriculum, the program, and the physical facilities need to take advantage of this integration.

PROCEDURES FOR DETERMINING FACILITY DESIGN

Any change in the physical structure of a facility is costly in time, energy, and money. In today's climate of reduced budgets, the general public needs to be assured that tax dollars are being spent wisely. Construction in technology education programs are no exception. Before any physical changes can take place, a variety of people need to give considerable thought to the question, "what, if any, needed changes should take place to support the curriculum?" This preplanning process should begin long before construction contracts are formalized. As reported by Polette (1993) the following list of four assumptions should be addressed before any actual facility planning begins.

1) All program faculty have reached a consensus on the philosophical base for the technology education program.

2) A curriculum plan has been reviewed and established for the construction in technology education program.

3) The administration supports the proposed curriculum and facility changes or, at least, has an open mind to allow change to take place.

4) There is a funding base to support facility renovation, remodeling, or new construction.

Without a clear understanding of the above factors and the support of the various participants in the educational setting, the success of any facility renovation will meet with a variety of road blocks and probably not be successful.

The facility needs to be viewed as: (1) a carefully planned educational resource that supports the goals and overall mission of the educational program, and (2) an integral part of the total instructional package provided by the educational system. To reach this goal, it is suggested that the teachers, school administrators, board members, and contractors follow a set procedure to avoid missing key details. This section outlines a model for arriving at a well-designed facility which will meet the educational needs of the student, Figure 9–2.

Determining Needs

The facility planning process must begin by establishing a philosophical base followed by a well-thought-out curriculum. Next, an assessment of current equipment and facilities is conducted to determine what existing equipment and space can be used for the new program. This information, coupled with the needs of the proposed program, can then be analyzed to determine if new or additional equipment and space are required to meet the goals of the new program or can these needs be met by making minor adjustments in the existing laboratory and its associated equipment (Rooke, 1988). Although this is not a simple process, it is essential to identify needs for the new construction program.

The Needs-Assessment Team

The needs-assessment team should include instructors in the technology education program, an administrator, advisory committee members, and student representatives. Consultants may be called in as the needs-assessment team attempts to gather data and requires specific information, space requirements, or equipment needs. The needs-assessment team relies on the previously agreed upon philosophy, program mission statement, and the selected curriculum. Regardless of how appropriate it would be to have a particular piece of equipment or additional space, if they do not meet the objectives of the program, they cannot be justified. Each objective is identified and then paired with activities that allow students to effectively meet the objective. Then, based on the selected activities, the specific pieces of equipment and space requirements are determined. The equipment selected determines the utilities needed, specific space requirements for

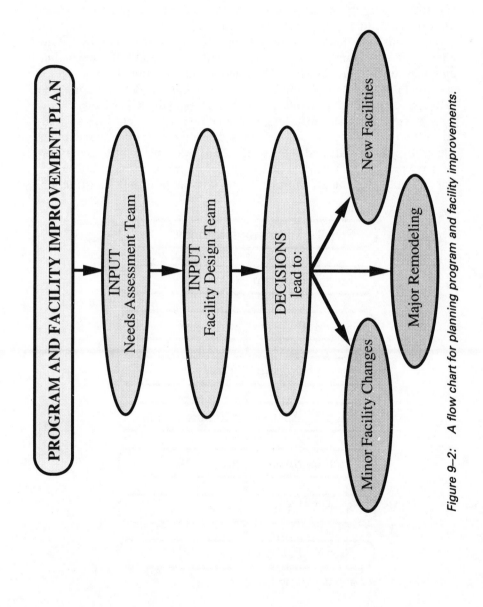

Figure 9–2: A flow chart for planning program and facility improvements.

activities, and storage for materials, teaching supplies, models, and equipment. See Figure 9–3.

Planners need to look carefully at the curriculum to identify multiple activities which are conducted simultaneously and determine if several activities require the same space. This may require more space or rescheduling activities to allow for better utilization of the facility.

The needs-assessment team makes one of the following decisions: (1) the construction program remains in the current facility and little remodeling and adjustments are needed, (2) remodel the existing facility, or (3) construct new facilities. As this decision is made, it may be appropriate to call upon outside consultants to help the team do a feasibility study. Architects, engineers, contractors, upper administration, etc., are all valuable resources. On the university campus, consultants are often found on the physical facilities staff and their services are available at little cost. Consultants provide information and help determine possible problems or opportunities that impact the cost of the renovation or new construction.

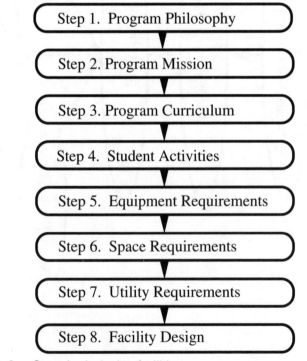

Figure 9–3:   Steps in designing facilities.

Another factor that the needs-assessment team must consider is the educational effectiveness of the current facility. It is possible that the facility is adequate, but has a serious limitation that makes it unsatisfactory. This may be due to limited or no access to the outside, limited access for handicapped students, or noise generated in the facility disrupts students in nearby classrooms.

The Design Team

The design team is usually headed by an architect and a team from that firm. Other important players include faculty members, administrators, engineers, and other specialized consultants. The design team's major function is to design a facility that will meet the goals and objectives established by the needs-assessment team. This requires careful coordination between teams as the design moves from the initial concept to the actual design.

Analyzing The Resources

Planning the construction technology education laboratory requires combined efforts of a variety of resources. People are perhaps the most critical of these resources. Cooperative effort by all individuals associated with the planning of the new or remodeled facility is required to ensure that the final plan meets original expectations. Data needs to be collected, analyzed, and evaluated to assure that the finished facility will, in fact, meet the educational goals of the program in an efficient manner. Finally, financial support must be established to pay for constructing and equipping of the facility.

People. All members of the needs-assessment and design teams must fully understand and support the philosophy, the specific mission of the program, and the curriculum design. Once this foundation is in place, both teams move forward by carefully reviewing the existing program, and facility and planning for the future. Once the needs-assessment is complete, the data can assist the design team with their tasks. The eventual output of the design team is a description of the general spaces and the relationship of each space to the others. A bubble diagram or conceptual model, such as shown in Figure 9–4, is often used.

Data. The needs-assessment and design teams rely on data such as enrollment trends, potential funding, educational trends, square footage recommendations, and the relationship of the construction program to other school offerings. Armed with this data and the curriculum design, the planning team can clearly see what should be included in the facility.

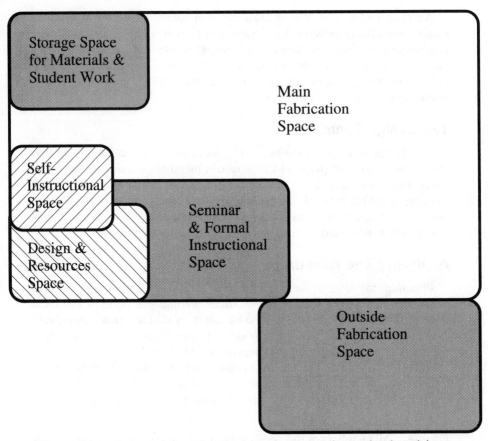

Figure 9–4: *Conceptual model of a typical construction technology labora-tory.*

Financial Arrangements. Available resources must be identified and secured before any extensive remodeling or new construction is started. While there may be a few dollars in the departmental budget to complete minor remodeling, there is seldom enough money to do major remodeling. Major remodeling requires that the departmental administration work with upper administration to seek other financial resources to fund the project. This may be a one-time allocation to the department or involve approvals from state government. In either case, the design team must convince decision makers that the proposed construction is needed, and is in the best interest of the students, program, and school. In times of limited financial resources, these presentations must be professionally done to be successful. It must describe the school's mission and generate the idea that without the proposed construction, the students will not receive adequate education to achieve success in their chosen career.

Scheduling. Scheduling of the project is critical to successful completion. A schedule that causes the least disruption to classes must be established early. Approvals may take several months on relatively simple projects and perhaps years when a new facility is built. A realistic schedule includes the time needed to complete the curriculum design, needs-assessment, and selection of specialized learning activities. If the schedule is not well thought out, it may endanger the entire project.

SPACE CONCEPTS

Design/Resource Space

Design/resource space, where the students have access to local resources, and is a critical component of the construction technology facility. The space should be isolated from the main construction laboratory to limit the infiltration of dust and noise. This space may be incorporated into a resource room that is shared with other clusters to reduce the costs. At lower-grade levels, the design/resource facility must be carefully monitored and should incorporate the classroom. The design/resource space must be large enough to accommodate a variety of media and, if possible, provide for future expansion. Reference books, videotapes, computers, modems, a variety of software, telephone lines, drawing, and sketching tables are also housed within the space. Computer software, videos, and other media used by either the student or the teacher may also be stored there.

Seminar/Formal Instructional Space

Seminar/formal instructional space generally includes the classroom where traditional instructional functions are conducted. An optimum size would seat 20–24 students. Today most instructors prefer tables and chairs rather than the traditional-desk arrangements. Tables provide more flexibility for a variety of activities and can be arranged in a variety of patterns to facilitate different instructional strategies. Computer work stations can be located along the walls on movable cabinets or tables. The seminar/formal instructional space should be located conveniently close to the main construction laboratory so that materials and equipment can be moved between them. It is desirable to have glass partitions between the spaces so visual contact is possible.

Storage Spaces

Space for storing materials and student work should be isolated from the main construction laboratory yet conveniently close. Storage spaces in the laboratory are difficult to manage and to maintain security. Material storage space should be large enough for a variety of construction activities. A minimum length would be 20 feet with horizontal storage racks as well as storage for sheet material with adequate room for several individuals to function in the storage room at the same time. Lockable storage cabinets should also be provided to store small items that are easily lost. A special cabinet must be provided for safe storage of flammable liquids.

Storage spaces for student work should include approximately one cubic yard for models of wall construction, concrete samples, or prepared samples to be tested. Larger student projects require outside storage or short-term storage in the laboratory itself. Well-designed storage spaces enhance the overall maintenance and neat appearance of the facility and projects a professional image.

Fabrication Space

The fabrication space in a construction laboratory needs to be large enough to accommodate a wide variety of construction activities. These activities include experiences such as mixing and placing of concrete, building concrete wall sections, conventional wood framing, the fabrication of small buildings, and structures, and testing materials. The laboratory needs high ceilings and a large overhead door to the outside, for the handling of materials and completed activities. Some activities can be conducted outdoors in a space several times larger than the interior laboratory. This requires easy access to tools, equipment, and utilities, and a large concrete slab. Both the indoor and outdoor laboratory space should have flexible utility components to provide for moving equipment and its efficient use. The laboratory floor area should allow for the rearrangement of stationary equipment and benches to maximize its efficient use. If the construction laboratory is shared with manufacturing, it may be necessary to rearrange the laboratory between terms in order to convert it from a manufacturing to a construction facility. For details of the recommended space requirements for the construction laboratory, review the CTTE Monograph 13, *Planning Technology Teacher Education Learning Environments* (Polette, 1991).

SPECIAL CONSIDERATIONS OF THE CONSTRUCTION LABORATORY

Rather than provide specific details of a particular construction laboratory, which is available in a variety of other publications, this section will outline features unique to a construction program.

Security

The management of tools, materials, and student work in a construction laboratory requires a security system that is both effective and easily operated. While there are many solutions to security problems, most will center on well-organized procedures with well-designed storage spaces for tools, materials, and projects.

The possibility that the facility will be used during evening hours or by other instructors tends to complicate the control of the laboratory. A departmental or school policy needs to be developed early so potential problems can be minimized before larger problems develop. This security policy might include such things as the controlled distribution of keys for the general laboratory, special cabinets for materials and computer software, locks on computers, and locks on certain pieces of equipment that should not be operated by inexperienced people.

The construction laboratory may require outside storage of materials and student work which will require some control via fences, gates, or walls. Whatever systems are installed, they must fit within the overall plan of the facility and not hinder instructional activities or the educational program of the school or university.

General Safety

The safety of the students and the instructor must be a prime consideration. Construction activities are hazardous because tools, equipment, and materials are used and overhead work is common. The wearing of hard hats and instruction on the proper use of scaffolding, ladders, and railings is essential.

A well-organized construction program considers safety as an integral part of the curriculum and the laboratory instruction. The attitude of the student worker plays a major role in a safe working environment. Attitude relates directly to the instructional program and proper maintenance of the construction laboratory. If the laboratory is not well kept it sends a signal to students that laboratory organization and proper care of materials and tools is not an important aspect of the construction industry.

Even in the best organized and operated construction program, there is a potential for accidents. The laboratory must have proper equipment to deal with these emergencies. Adequate fire extinguishers of the proper type, a cabinet for safety glasses for visitors or students, and a first-aid kit suitable for responding to typical injuries must be available. All safety items must be located in clearly identified and easily accessible locations.

Each laboratory space should have a clearly posted evacuation plan that provides routes and assembly points in case of a fire or natural disaster. In all states, public buildings must meet certain fire requirements with the plans and facilities approved by the fire marshal's office to assure compliance with building codes. Further, most states have safety plans on how to dispose of certain toxic substances. The design team must check with state and local building officials early during the planning process to insure that the completed facility will meet all safety codes.

Noise And Climate Control

A properly designed construction facility provides a climate that is conducive to learning. Proper temperature, humidity control, and noise reduction should be effective. Noise not only distracts from the quality learning environment but also can cause hearing loss; especially to teachers who experience it throughout the entire day. Many construction activities require the use of noisy power equipment and care should be taken in its selection. The laboratory itself can be fitted with sound-absorbing treatments to the walls and ceiling which can do much to reduce the reflected sound.

Temperature control is important for an efficient learning environment. In fact, if the temperature is out of comfort limits it can become a major safety factor. Heating and air-conditioning equipment needs to be properly sized and should be designed by a professional engineer. Systems may need to be oversized to accommodate for the opening of large overhead doors.

As costs continue to rise, there is a real possibility that year-around school will be implemented. If the facility is to be used during the summer months, air-handling equipment must be adequate to assure a comfortable learning environment. Consider dust-control systems when designing air-handling equipment. Also be concerned about humidity. High humidity can cause tools and equipment to rust. Also, humidity can be so low that static electricity charges build up to a point that it may damage sensitive computer equipment.

Utility Systems

Because of the large demand for electrical circuits, telephone and computer lines, hot and cold water, and compressed gasses, designers must specify utility systems to handle future needs. Installation of utilities is relatively inexpensive during the initial construction phase but becomes much more costly if it is installed or moved at a later date.

The location of specific pieces of large equipment require connections to a utility system. These connections must be carefully analyzed to assure that specific pieces of stationary equipment will not present future problems. It may be desirable to establish several locations where this equipment can be moved to enhance laboratory flexibility.

Accessibility For Special-Needs Students

The right of special-needs students is assured by the Education for All Handicapped Children Act, PL 94–142 (1975). This legislation requires that all new or remodeled facilities provide for the accessibility of handicapped individuals. This has considerable impact on the design of laboratories. Doors, curbs, work stations, or other artificial barriers which inhibit accessibility need to be taken into account so these students can participate fully in educational activities. Although these requirements are not difficult to implement in new facility designs, they do need to be addressed early in the design of renovated facilities. For more detailed information on designing facilities to meet these codes, review the International Technology Education Association's professional monograph *Making Industrial Education Facilities Accessible to the Physically Disabled* (Shackelford & Henak, 1982).

SUMMARY

Construction technology facilities provide the environment for instruction in construction in technology education. These facilities vary depending upon the educational level of the program. At the elementary level, it is appropriate to integrate the construction activities into the standard classroom. At the middle school level, the construction program is most appropriately integrated into the total technology education program and does not require a separate laboratory. The high school program may require separate facilities depending on the size and scope of the offerings. At the teacher education level, the construction laboratory can be separate, or housed in conjunction with the manufacturing program. At all levels, the laboratory must provide a safe and stimulating environment that is flexible and easily adjusted to future needs of the total technology education program.

A well-thought-out philosophy and curriculum design must be in place prior to any changes or additions to the construction technology facility. New facilities or the remodeling of existing facilities requires the thoughtful input by a needs-assessment team who assesses the curriculum and learning activities, which require a variety of tools, materials, equipment, and supplies. Based on the findings of the needs-assessment team, a design team develops the actual layout of the new or remodeled facilities. During the design process, care is exercised to assure the facility will have adequate design/resource, seminar/formal instructional, storage, and fabrication spaces. Accessibility for special-needs students and adequate noise and climate control must also be addressed.

REFERENCES

Bame, A.E. (1986). Middle/junior high technology education. In R.E. Jones & J.R. Wright (Eds.), *Implementing technology education* (pp. 70–94). Encino, CA: Glencoe.

Braybrook, S. (Ed.). (1986). *Design for research: Principles of laboratory.* New York: Wiley Interscience.

Broadwell, M.M. (1979). Classroom instruction. In R.L. Craig (Ed.), *Training and development handbook: A guide to human resource development* (pp. 33–1 through 33–13). New York: McGraw-Hill.

Brown, R. (1990). Establishing the communication teaching and learning environment. In J.A. Liedtke (Ed.), *Communication Technology Education* (pp. 139–164). Encino, CA: Glencoe.

Cummings, P.L., Jensen, M., & Todd. R. (1987). Facilities for technology education. *The Technology Teacher. 46*(7), 7–10.

Daiber, R.A., & LaClair, T.D. (1986). High school technology education. In R.E. Jones & J.R. Wright (Eds.), *Implementing technology education* (pp. 95–137). Encino, CA: Glencoe.

Doyle, M. (1982). Learning environments for design and technology. *Technology, Innovation, and Entrepreneurship for Students. 4*(4), (pp. 43–47).

Erekson, T.L. (1981). *Accessibility to laboratories and equipment for the physically handicapped: A handbook for vocational education personnel.* Springfield, IL: Illinois State Board of Education, Division of Vocational and Technical Education.

Gemmill, P.R. (1979). Industrial arts laboratory facilities: Past, present and future. In G.E. Martin (Ed.), *Industrial arts education retrospect, prospect* (pp. 86–101). Bloomington, IL: McKnight.

Gould, B.P. (1986). Facilities programming. In S. Braybrook (Ed.), *Design for research; principles of laboratory architecture* (pp. 18–48). New York: Wiley Interscience.

Hales, J.A., & Snyder, J.F. (1981). *Jackson's Mill industrial arts curriculum theory.* Charleston, WV: West Virginia Board of Education.

Helsel, L.D., & Jones, R.E. (1986). Undergraduate technology education: The technical sequence. In R.E. Jones & J.R. Wright (Eds.), *Implementing technology education* (pp. 171–200). Encino, CA: Glencoe.

Henak, R. (Ed.). (1989). *Elements and structure for a model undergraduate technology teacher education program.* Council on Technology Teacher Education Monograph 11, Reston, VA: International Technology Education Association.

Huss, W.E. (1959). Principles of laboratory planning. In R.K. Nair (Ed.), *Planning industrial arts facilities* (pp. 48–65). Bloomington, IL: McKnight.

James, R.W., & Alcorn, P.A. (1991). *A guide to facilities planning.* Englewood, NJ: Prentice Hall.

Johnson, F.M. (1979). The organization and utilization of a planning committee. In D.G. Fox (Ed.), *Design of biomedical research facilities: Proceedings of a cancer research safety symposium,* Report No. 81–2305 (pp. 3–16). Bethesda, MD: National Institutes of Health.

Kemp, W.H., & Schwaller, A.E. (1988). Introduction to instructional strategies. In W.H. Kemp & A.E. Schwaller (Eds.), *Instructional strategies for technology education* (pp. 16–34). Mission Hills, CA: Glencoe.

Lauda, D.P., & McCrory, D.L. (1986). A rationale for technology education. In R.E. Jones & J.R. Wright (Eds.), *Implementing technology education* (pp. 15–46). Encino, CA: Glencoe Publishers.

Lewis, H.F. (1951). Site selection, design, and construction. In H.S. Coleman (Ed.), *Laboratory design* (pp. 75–79). New York: Reinhold.

McHaney, L.J., & Berhardt, J. (1988). Technology programs: The Woodlands, TX. *The Technology Teacher, 48*(1), 11–16.

Polette, D.L. (Ed.). (1991). *Planning technology teacher education learning environments.* Council on Technology Teacher Education Monograph 13, Reston, VA: International Technology Education Association.

Polette, D.L. (1993). Facilities for teaching manufacturing. In R.D. Seymour & R. Shackelford (Ed.), *Manufacturing in technology education* (pp. 189–214). Encino, CA: Glencoe.

Rokke, D.L. (1988). *An overview of technology laboratory design.* Paper presented at the annual conference of the American Vocational Association, St. Louis, MO.

Rokusek, H.J., & Israel, E.N. (1988). Twenty-five years of change (1963–1988) and its effect on industrial teacher education administrators. In D.L. Householder (Ed.), *Industrial teacher education in transition: Proceedings of the 75th Mississippi Valley Industrial Teacher Education Conference* (pp. 219–244). College Station, TX: Mississippi Valley Industrial Teacher Education Conference.

Savage, E., & Sterry, L. (Eds.). (1990). *A conceptual framework for technology education.* Reston, VA: International Technology Education Association.

Seal, M.R., & Goltz, H.A. (1975). Planning principles. In D.E. Moon (Ed.), *A guide to the planning of industrial arts facilities* (pp. 55–64). Bloomington, IL: McKnight.

Shackelford, R., & Henak, R. (1982). *Making industrial education facilities accessible to the physically disabled.* Reston, VA: International Technology Education Association.

10

Assessment Of Learning And Instruction Of Construction In Technology Programs

Scott D. Johnson, Ph.D.
Department of Vocational & Technical Education
University of Illinois at Urbana-Champaign, Champaign, IL

Assessment of learning and instruction is a critical component of quality education. Learning assessments determine the degree to which students attain educational goals and objectives. If assessments reveal that students are failing to reach the learning objectives, alternative strategies for instruction can be developed and implemented. Instructional assessments provide information to instructors so they can reflect on and improve their instructional technique. Without thoughtful and focused assessment, the quality of learning and instruction will be greatly diminished.

Current practice in educational assessment is often based on customs that have been handed down by each generation of instructors. By being constantly assessed through their primary and secondary school experiences, students become very experienced and knowledgeable about assessment. When students progress into teacher education programs, they carry with them a set of beliefs about assessment that are often reinforced in their teacher preparation courses. As a result, the methods of assessment used by former instructors tend to influence the beliefs and practices of future teachers.

Unfortunately, much of the assessment activity that takes place in American education is deficient in many ways. The purpose of this chapter is to examine current methods of educational assessment, to identify their innate weaknesses, and to describe contemporary approaches to assessment that can improve the quality of learning and instruction in technology education. While much of the content of this chapter pertains to assessment in construction technology programs, the concepts and methods that are presented are equally appropriate for all areas of technology education.

PROBLEMS WITH CURRENT METHODS OF ASSESSMENT

The predominant methods of assessment in education leave much to be desired. The most troubling of the weaknesses are the overemphasis on testing, the focus on specific skills rather than holistic performance, and the focus on results rather than processes.

Overemphasis On Conventional Testing

Tests began to have a significant impact on American education when the National Assessment of Educational Progress (NAEP) was founded by Congress in 1969. NAEP's role was to determine trends in education and report those trends to Congress. With the need to periodically conduct national assessments of 9, 13, and 17 year olds came an overreliance on multiple-choice tests. The increased emphasis on multiple-choice tests was primarily due to their economy, efficiency, and objectivity in administration and scoring (Frederiksen, 1990).

As the prevalence of testing in the schools increased, educators became more dissatisfied with the notion of testing. Many educators now believe that tests can have a negative effect on teaching and learning (Kirst, 1991). The reasons for the dissatisfaction with tests include their tendency to:

1) Promote memorization of facts rather than understanding of concepts (Collins, 1990; Linn, 1991).

2) Increase drill and practice activities in classrooms.

3) Neglect intellectual processes such as problem solving and decision making and provide little information regarding the misconceptions students may hold (Frederiksen & White, 1990).

4) Lead students to believe that there is always one right answer, that problems are well-structured, that guessing is inappropriate, and that the right answers always reside in the head of the teacher (Collins, 1990; Kirst, 1991).

5) Lead to a narrowing of the curriculum as teachers "teach to the test" and students learn what the test will measure (Frederiksen, 1990; Kirst, 1991).

Focus On Specific Skills Rather Than Real-World Performance

Assessments of learning tend to focus on students' ability to perform specific skills and pay little attention to actual performance on real-world tasks (McGaghie, 1991). This tendency is driven by curriculum developers who often develop long lists of tasks that students should be able to do upon successful completion of a course. These task lists become the focus of learning assessments as students are tested to identify the tasks they can and cannot complete.

Current research in cognitive psychology suggests that teaching should emphasize more than specific-skill development because instruction that focuses on specific skills tends to result in the development of inert knowledge; that is, knowledge that students possess but cannot use (Whitehead, 1929). Assessing the ability of students to perform such isolated tasks or skills provides instructors with little evaluative information. A more fruitful approach is to assess students as they apply their knowledge and skills to solve problems that involve combinations of specific skills.

Focus On Results Rather Than Processes

It is common for instructors to focus their assessment activity on end products. Final exams, research reports, oral presentations, construction budgets, materials lists, and construction models are all examples of the end products that are commonly used to assess student knowledge and skills. While end products do provide a valuable source of evaluative information, they do not reveal the processes used by the student to complete them. Even if the end product looks good and meets the expectations of the instructor, it is possible that faulty procedures were followed. Instructors need to expand their assessments of learning to include both the processes used and the final products that are created.

ADOPTING AN ASSESSMENT CULTURE

A recent review of educational research on assessment argues for a change in assessment philosophy (Wolf, Bixby, Glenn, & Gardner, 1991). Wolf and colleagues contend that the current emphasis on testing has created a culture that views learning and assessment in ways that are unproductive for maintaining quality education. To make the change to an assessment culture, Wolf et al. (1991) advocated adoption of the following principles:

1) Educators need to view assessment as an occasion for learning rather than testing. Proper approaches should not only assess the outcomes of learning, but also create the opportunity to improve learning and instruction.

2) Assessment should be formative and ongoing rather than viewed as a "point in time" evaluation of knowledge. As this principle gains acceptance and becomes implemented, the need for final exams may disappear (Frederiksen, 1990).

3) A criterion-referenced philosophy of evaluation is needed which leads to assessment focused on accomplishment rather than rank.

4) Learning is more than an accumulation of knowledge that is verified through recognition and recall tests. An assessment culture views learning as a constructive process that is verified by application, use, and transfer of knowledge in realistic contexts. Tests should emphasize learning and thinking, require generation as well as selection, and be an ongoing occurrence during learning rather than serving as exit points for instruction (Collins, 1990).

5) In a testing culture, correctness is desired, simple and low levels of understanding are commonly evaluated, and test content is often determined by what is easy to score. In a reformed assessment culture the process of performance is valued beyond simple correctness, the content of assessment is based on what students need to know and do rather than on what is easy to score, and there are exhibitions of invention, transfer, and inquiry.

As technology educators begin the transition to an assessment culture, traditional methods of assessment will change. Rather than relying on multiple-choice and essay tests as a primary method of assessing student learning, instructors need to endorse ongoing assessment through multiple- and authentic-assessment tactics. Because many of the learning activities encountered in construction technology cannot be quantified on paper-and-pencil tests, assessments of learning must also focus on the direct assessment of practical skills. These assessments include traditional techniques such as paper-and-pencil exams, interviews, and classroom observations as well as innovative approaches such as situational tests, portfolio reviews, and performance exams. In addition, increased attention should be given to assessing professional and personal qualities such as honesty, judgment, work habits, maturity, psychological stability, and adaptive capabilities (McGaghie, 1991).

Assessment in construction technology can be divided into two major categories; assessment of products and assessment of processes. Assessment of products includes the evaluation of student's writing samples, learning activities, homework, and major projects that involve planning and constructing a product. Assessment of processes involves instructor observations of students working on projects, interviews of students, and paper-and-pencil tests covering both basic skills and construction-related information.

ASSESSMENT OF PRODUCTS

Assessment of products provides instructors with an indication of student's ability to complete construction-related tasks. This form of assessment is primarily concerned with end products, that is, with the products that are ultimately created by students.

Project Evaluation

Whenever construction work is completed in the field, it is subjected to a thorough inspection. This inspection can involve public officials, the contractor, the design team, financial representatives, suppliers, and the owner. The final inspection involves examination of every part of the job, in detail, to see if the work has been completed according to the construction contract. The structural integrity of the project, as well as the proper operation of all mechanical and electrical systems are closely examined. A list of defects, called a "punch list," is created during a final inspection. The punch list is used to communicate to the contractor what needs to be corrected in order to complete the job.

The evaluation of student projects in construction technology education should be based on the inspection process that occurs in the field. Inspection teams involving students, community representatives, and the instructor should participate in the inspection. This inspection should be similar to a real inspection and focus on the quality of the materials used, the appropriateness of the construction methods, and the overall quality of work (Henak, 1993, pp. 301–304). Students should be made aware of the expectations for quality before they begin work and these criteria should be the focus of evaluations of construction projects. A sample punch list that could be used to inspect a final construction project is shown in Figure 10–1.

Following an inspection, contractors are given the opportunity to correct any deficiencies in their work. This practice should also be used in the construction technology education classes. Following the final inspection,

---

PUNCH LIST

TO: Contractor: _____ _____ Date: _____

FROM: Designer _____ Owner _____ Building Inspector _____ Lender _____ Supplier _____

SUBJECT: Inspection and Punch List for _____ _____
(Project Name)

The items on the list below were found to be unsatisfactory. They must be corrected before the Certificate of Completion is signed.

1. _____

2. _____

3. _____

4. _____

5. _____

6. _____

7. _____

8. _____

9. _____

10. _____

INSPECTOR'S SIGNATURE: _____ _____ Position: _____

Instructions: Use the inspection checklist below. It will help you to check each item. You may add items to be checked as needed. The items that pass inspection are checked under the S (Satisfactory). Place a check under the U (Unsatisfactory) for those that do not pass. List the unsatisfactory items on the Punch List

ITEMS TO INSPECT	S	U
1. Is the roofing put on according to the specifications and drawings?		
2. Do the exterior wall coverings meet material and quality specifications?		
3. Are all cracks sealed?		
4. Was the specified paint used and applied correctly?		
5. Were the specified wall, ceiling, and floor materials correctly installed?		
6. Were electrical circuits and devices installed according to specifications?		
7. Were the specified plumbing materials used and correctly installed?		
8. 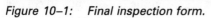 (Add or subtract items as needed.)		

Figure 10–1:   Final inspection form.

students should be given the opportunity to correct any defects in their work. Grades can then be calculated based on the results of the final inspection and the student's ability to make corrections.

Comprehensive Portfolio

In construction technology, it is usually difficult for students to hand in the results of their work because construction activities result in large and immovable products. Because of the unique nature of construction technology education products, it is desirable to have students create a comprehensive construction portfolio that records their progress in the course. The contents of a construction portfolio can vary although both the instructor and the students should be involved in selecting what is to be included. A portfolio for a construction technology education course could contain the following:

Reflective Journal. Students should be expected to maintain a journal that describes their thoughts and experiences in the course. The journal will serve as a record of the changes that have occurred in the student's knowledge of, and attitudes toward the construction field. The journal can also be used to enhance the writing abilities of students which is a skill that is increasingly desired by employers. Instructors should periodically read the journals to keep informed of their student's concerns and difficulties in the course.

Videotapes. Because construction is a physical trade that involves competencies with tools and materials, a videotape of the student engaged in construction work can provide an accurate record of the student's abilities. One videotape can be used throughout the course to record students in action. This videotape can then be used to show the changes that have occurred in a student's abilities throughout the course.

Photographs. Pictures of construction products should also be included in the portfolio. These pictures can show the overall product and specific details within the project. For example, if a class was involved in the construction of a storage shed, a photograph of the completed shed should be included. In addition, close-up photographs of the framing, sheathing, and roofing would be needed to show the quality of specific aspects of the product.

Diagrams, Drawings, and Lists. All drawings, sketches, blueprints, and lists of materials created by the student should be included in the portfolio. These written materials serve as a record of the student's planning and organizational skills.

Written Work. All of the written materials completed throughout the course should be included in the portfolio. This includes copies of tests, quizzes, and homework assignments.

Computer Programs. If students use or develop computer programs that support construction activities, they should also be included in the portfolio. Examples include printouts created from budgeting or construction-planning software.

The use of a comprehensive portfolio provides the instructor with a thorough record of the student's progress and performance in the course. Instructors should periodically review these portfolios to ensure that students are actually maintaining them, to gain an understanding of student's strengths and weaknesses, and to determine and justify the grades that are given. These portfolios are also effective communication tools for displaying student achievement to parents and administrators.

ASSESSMENT OF PROCESSES AND UNDERSTANDING

While assessments of products can provide an indication of students' ability to complete construction technology education projects, further assessments are needed to evaluate the quality of the processes students use and their depth of understanding. Assessment of processes and understanding are really performance assessments. Performance assessments can involve using paper-and-pencil examinations, directly questioning students, or observing them as they complete tasks.

Paper-And-Pencil Testing

The instructor-developed tests used in construction technology courses are criterion-referenced examinations. Criterion-referenced tests assess what students know, determine how successful they have been in mastering the course content, and identify the course content that students do not know so instruction can be altered to emphasize those areas.

Instructors often misinterpret criterion-referenced scores. Just because two individuals receive identical scores on a criterion-referenced test, it does not necessarily mean that they have mastered the identical content. For example, on a test designed to assess student knowledge of municipal building codes, one student might correctly answer all questions about electrical codes but incorrectly answer structural-code questions. A second student, who received the same overall score on the test, might do well on the structural codes but do poorly on the electrical-code questions. In this example, each student could have received the same overall score on the test but have significantly different knowledge of the content. A careful analysis of individual test scores is needed to identify specific deficiencies for each student.

A major problem with instructor-developed tests is that they typically involve assessment of only the lower levels of learning. Consider the following questions taken from popular construction technology textbooks:

1) A bill of materials lists the raw materials and parts needed to produce an item. (T or F)

2) House plans include a written list of materials that is known as the _____.

3) Concrete walks, steps, and patios are poured by a worker known as the _____.

4) What common silvery metal does not corrode easily?

5) A "Square" of shingles will cover _____ sq. ft. of roof.

Each of the above questions requires students to simply recall information that was previously learned. By asking these types of questions, instructors learn very little about a student's depth of understanding or their ability to cognitively process information. While it is important for students to learn and remember certain facts and concepts, it is also important for students to develop the cognitive skills needed for higher levels of performance. Consider the problem of transferability of knowledge. Will the facts and concepts that are learned today be as relevant and accurate tomorrow? The growth of knowledge in recent years has been phenomenal. Recent changes in technology have had a dramatic impact on the construction industry. Many of the tools, materials, and processes that were in common use five years ago are virtually obsolete at this time.

In view of the changes in technology, technology educators need to place greater emphasis on higher-level cognitive process skills because these are the skills that are transferable. Instead of leaving a construction technology course with a mind filled with facts and concepts that may not be relevant in the near future, students must have the cognitive process skills that are needed to gather, interpret, analyze, synthesize, and evaluate information. In other words, students must have the ability to "learn how to learn." It is important for technology educators to emphasize the learning experiences and activities that develop these transferable skills at the higher dimension of knowledge and also plan evaluations that measure these skills.

Preparing a Table of Specifications. To insure that teacher-developed tests assess both low and high levels of cognition, instructors should begin their test development process by creating a table of specifications. Few contractors begin a project without a plan. In order to prevent costly mistakes, contractors use a drawing to help them make construction decisions. In the same fashion, the test developer who begins writing test items without prior planning will likely develop a weak test. In place of a construction drawing, test developers use a table of specifications. This table provides the information needed to guide the test-development process.

The first step in preparing a table of specifications is to prepare an outline of both the instructional objectives and the subject matter covered in the course. Next, the table of specifications may be prepared. A table of specifications, shown in Figure 10–2, ensures that the test question selections are based on the amount of instructional emphasis placed on the content. Information for the "% of emphasis in curriculum" column is obtained from lesson plans. The second and third columns represent the number of content and process questions to include. Content questions are those that require students to use low-level cognitive skills such as recall and recognition in order to correctly provide answers. Process questions are those that require students to use high-level cognitive skills such as problem solving, analysis, and synthesis. The final column is calculated by multiplying the number of points on the test by the percent of emphasis for each particular topic. For example, if a 50 point test is desired and if 10% of the class time was spent covering rough framing, then 10% of the 50 points, or 5 total points, should relate to that topic.

Writing Test Items. Writing quality-test items is a difficult task for even the most experienced test writers. Even if extreme care is taken to write what appears to be quality-test items, the results may be ambiguous to the reader. Poorly written test items might also provide clues to good guessers and they may not even test for the content or skills for which they were designed. Numerous texts have been written that describe the advantages and disadvantages of the various test-item formats and provide guidelines that can help ensure that reliable and valid items will be written. See Erickson and Wentling (1988) for an example of a text that thoroughly covers test-creation procedures.

While it is beyond the scope of this chapter to cover the process of writing test items, there are several key points that must be made. Test-item formats such as true-false, multiple-choice, matching, short-answer, and essay are all used for different purposes. For example, true-false, multiple-choice, and matching items require students to select or recognize the most appropriate response from a list of several possible answers. Completion, short-answer,

Content Area Taken from Learning Objectives	% of Emphasis in Curriculum	Low-Level Knowledge Items	High-Level Knowledge Items	Points on Exam
Selecting the Site - Criteria for selecting sites - Surveying a site	20%	6	4	10 points
Acquiring the Site - Examining site history - Soil testing - Obtaining records - Titles and surveys - Changing ownership	30%	6	9	15 points
Clearing the Site - Preparing to clear the site - Major clearing operations	15%	5	3	8 points
Locating the Structure - Types of surveys - Conducting surveys	25%	4	8	12 points
Mass Excavation - Trimming and shaping - Stabilizing	10%	2	3	5 points
Totals	100%	23 points	27 points	50 points

Figure 10–2: Table of specifications for unit exam covering site preparation.

and essay-type questions require students to provide their own answers to questions and problems so students must recall or reconstruct information from memory in order to respond to the test item. Figure 10–3 shows how test items covering similar content can be written in formats that test for very different things.

True-false, multiple-choice, and matching items are popular with instructors because they can be graded quickly. However, these types of items are inherently weak because they encourage guessing. True-false items are especially guilty of this deficiency because a student who knows absolutely nothing about the material can reasonably expect to get about half the questions right. Therefore, true-false items are generally unreliable and should be avoided.

High quality multiple-choice items are effective for assessing student learning and allow a great amount of material to be covered in a short amount of time. However, when items are poorly written, even students with minimal knowledge can eliminate one or two of the choices, thus improving the odds of guessing the correct answer to 1:2 or 1:3. To help avoid this, all of the choices provided should be: (1) plausible, (2) about the same length, and (3) grammatically compatible with the question. Phrasing the choices so that the correct answer is consistently the longest or shortest item is also a giveaway; and questions that answer another question on the test should also be avoided.

Short-answer and essay items reduce the chances of guessing. Short-answer items are written to test for learning things such as factual knowledge and comprehension of principles. They are also useful for assessing the ability of students to perform cognitive skills such as the ability to identify and define concepts. Essay questions are written to assess the ability of students to organize, integrate, and evaluate knowledge. Answers given to essay items also reflect students' attitudes, creativity, and verbal fluency. Short-answer and essay items are much easier to write than multiple-choice items, but they take much longer to grade. They also tend to be more subjective than multiple-choice items which reduces the reliability of the test. Short-answer and essay items also take students longer to complete which limits the amount of content that can be covered on a test.

Short-answer and essay items can be used to evaluate various types of learning ranging from simple recall to decision making and problem solving. When used to assess the ability of students to recall information, these items are written in the form of questions or statements that require students to remember information that was presented earlier. When short-answer and essay items are used to assess the higher-level cognitive skills, items are written in the form of problems that must be solved by the student.

A typical **multiple-choice item** to test mathematical understanding:

> *The square footage of a home that measures 25' x 50' is*
> *a. 2 sq. ft.*
> *b. 1250 sq. ft.*
> *c. 2500 sq. ft.*
> *d. 75 sq. ft.*

NOTE: In this item all of the answers are the results of different mathematical calculations of the numbers 25 and 50. The selection of plausible distracters like these ensures that the student understands the process and reduces the possibility of correctly guessing the answer.

An **open-ended item** that assesses mathematical understanding:

> *What is the length of a rectangular house that is 25' wide and has 1250 sq. ft. of floor space?*

A **performance task** that assesses both conceptual understanding and creativity:

> *Sketch as many floor-plan configurations as you can that have the same square footage as a rectangular building that measures 25' x 50'. Show your dimensioning and calculations.*

A **performance task** that assesses conceptual understanding and inquiry skills:

> *Measure the square footage of several residences. Develop a table that shows the range of sizes of each of the room types. Determine an average size for each type of room.*

A **word problem** to assess conceptual and mathematical understanding, knowledge of basic residential design principles, and the ability to reflect on the creative process:

> *You have been hired to design a house that has 1250 sq. ft. of floor space. Your clients want the house to contain two bedrooms, a kitchen, a small dining area, a living room and a full bathroom. Sketch a floor plan that will meet this request. Be sure to include dimensions, room labels, and show your floor-space calculations. Explain how you arrived at the layout and the room sizes.*

An item for **student self-assessment**:

> *How well do you understand the process of calculating the square footage of a building? How do you think your understanding of this process could help you in the future?*

Figure 10–3: Sample test items (Adapted from Stenmark, 1989).

Possibly the biggest problem with short-answer and essay items occurs when scoring. Because it is virtually impossible to write every item so it has only one correct answer, the test scorer must consider alternative answers before the test is administered so objective scoring can take place. At the time the scoring takes place, the test scorer is often forced to make decisions regarding the correctness of various responses. This difficulty not only increases the length of time it takes to score these items, it also decreases the reliability of the total evaluation instrument.

Observations

Performance assessment also involves directly observing and questioning students as they perform real-construction tasks. The objective is to determine how well students are able to perform tasks and to identify the mistakes they make so additional instruction can be planned. This type of assessment is especially difficult because it tends to be both subjective and time-consuming.

To decrease the subjectivity involved in observing students and to speed up the evaluation process, most instructors rely on some form of rating scale to assess performance. As shown in Figure 10–4, rating scales divide the performance being observed into several subskills. The instructor then rates the quality of the student's performance on each of these subscales. In addition, instructors should attempt to make a holistic assessment of the student's overall performance. By doing this, the instructor will have an impression of the student's overall performance and specific assessments of individual skills. The results obtained from these rating scales can be used to assign grades, modify instruction, or provide corrective feedback to students.

Questioning Of Students

Instructors can gain a great understanding of a student's knowledge and cognitive processing ability by the deliberate use of questions. While questioning individual students can be very time-consuming, it does provide assessment information that is difficult to obtain by any other method. Through a series of questions that build upon the responses from prior questions, instructors can delve into the mind of the student to determine the quantity and quality of the student's knowledge. This verbal interaction can also illuminate a lack of knowledge and weaknesses in cognitive processing.

Several different types and levels of questions can be asked to assess understanding. These can be divided into low- and high-level questions and open and closed questions. Low-level questions are usually appropriate for

Performance Checklist

Student name: _____ _____

Task being observed: _____ _____

Circle the appropriate assessment that indicates the level of performance for each of the following criteria:

Performance Criteria	Level of Performance		
Demonstrates safe work habits.	Above Expectations	Meets Expectations	Below Expectations
Demonstrates proper work techniques.	Above Expectations	Meets Expectations	Below Expectations
Follows proper procedures.	Above Expectations	Meets Expectations	Below Expectations
Displays an organized plan of action.	Above Expectations	Meets Expectations	Below Expectations
Does accurate work.	Above Expectations	Meets Expectations	Below Expectations
Works efficiently.	Above Expectations	Meets Expectations	Below Expectations
Uses tools properly.	Above Expectations	Meets Expectations	Below Expectations
Overall quality of the performance.	Above Expectations	Meets Expectations	Below Expectations

Instructor comments regarding student's performance:

Figure 10–4: Sample performance checklist.

evaluating student comprehension, diagnosing their strengths and weaknesses, and reviewing or summarizing content. Generally, they only require students to recall or reiterate something that has been covered in a lesson. High-level questions are usually appropriate for encouraging students to think more deeply or critically, to problem solve, or to stimulate students to seek more information on their own. Closed questions are ones for which there are a limited number of acceptable answers, most of which can be anticipated by the instructor. Open questions are ones for which there are many acceptable answers, many of which will not be anticipated by the instructor. Examples of each of the types of questions include:

- **What is the plywood below the finished floor called?**
  This is a closed-type question because there is only one correct answer and is low level because it involves simple recall of a memorized fact.

- **What is a common material used for roofing?**
  This is an open-type question because there are several correct answers and is low level because it involves simple recall of memorized facts.

- **Based on the information on the drawing before you, would you say this structure is designed for residential or commercial use?**
  This is a closed-type question because there is only one correct answer and is high level because the student must use analytic thinking skills to determine the correct response.

- **What are some design changes that could make this home more energy-efficient?**
  This is an open-type question because there are many possible answers and is high level because the student must use analytic thinking skills to determine the correct response.

Most instructors are not very good at questioning. Fortunately, questioning is a skill that can and should be developed. The following hints can help improve questioning ability:

- Avoid questions that can be answered with a YES or NO unless you intend to follow with more questions.

- Avoid the "guess what I am thinking" type of question (20-question game).

- Use probing questions.

- Phrase questions so they are clear to students.

- Be prepared to break your question into simpler questions if students cannot answer correctly the first time.

- Be sure to allow "wait time" after asking a question.

- Be aware of your facial and body gestures as you ask/answer questions.

- Maintain eye contact with the student answering the question.

- Listen to student answers without interrupting.

- Repeat or paraphrase answers.

- Provide positive reinforcement for correct responses.

- Be tactful when dealing with incorrect answers. Students often do not appreciate cynicism or irony at their expense.

EVALUATION OF INSTRUCTION

Assessment in technology education involves more than the evaluation of student achievement. Assessment can also be used to evaluate the instructional process which can lead to more effective instruction in the future. Poor communication skills, low quality instructional materials, and ineffective instructional strategies are just a few examples of the weaknesses that an evaluation of instruction might identify.

The hardest part of instructional evaluation is making the commitment to conduct the evaluation. Evaluations of instruction are conducted to serve one major goal—to identify ways to improve instruction. This is a worthy goal and one which every instructor should attempt to reach.

Several strategies can be used to evaluate instruction. These include using tests and student questionnaires to assess instruction, being observed by peers, and using self-assessment through videotaping and reflection.

Using Tests To Assess Instruction

Tests measure more than student achievement; they also measure the effectiveness of the instructional process and can serve as a guide for improving instruction. In terms of instructional effectiveness, the most common use of test results is to see how many students missed a particular question. If most students get a question right, then the topic must have been effectively taught and learned. If a large number of students choose the same wrong answer, any number of factors could be at work. The question may have been phrased poorly or incorrect information may have been covered in class. If a wide variety of incorrect answers was given, the students were probably guessing. Instructors must be sensitive to all of these possibilities when grading a test.

Regardless of the source of the problem, when a large number of students miss a question or group of questions, it generally indicates that the material should be retaught. Reteaching is a difficult task because first the incorrect learning must be undone. When a point is being retaught, it is usually not sufficient to simply give the same lesson again. If it didn't work the first time, it probably won't work the second time either. Instead, try looking at the topic from a new viewpoint. An analysis of incorrect answers and discussions with the students might identify the misconceptions students must overcome and what new instructional approaches should be tried.

Using Questionnaires To Assess Instruction

Students' close and constant exposure to an instructor makes them an excellent source of evaluation information. The students may actually be the most qualified to evaluate instruction because they have firsthand knowledge of the effectiveness of the instructor, the effectiveness of the instructional methods, and quality of the instructional materials.

Several concerns exist regarding the use of students as evaluators of instruction. First, there is the concern that student evaluations of instructors become popularity contests. Second, students, as a group, are often whimsical, inconsistent, and unreliable judges of instructional effectiveness. While these are popular views among instructors, the research does not support these concerns. Students have been found to be discriminating judges of instructional quality and their ratings reflect more than popularity. Student evaluations have also been found to be reliable and valid. In fact, student evaluations appear to correlate positively with other measures of instructional effectiveness.

One of the best ways to obtain information about students' opinions of an instructor is question them about what they like in the course and what suggestions they would make for improving it. When conducting an opinion survey it is important to assure student confidentiality. It must be made clear to students that they are not expected to put their names on the questionnaire.

Questionnaires for gaining student opinions about an instructor typically consist of a set of open-ended questions. These questionnaires can be constructed so they require the students to either select an appropriate answer or provide their own. When the questionnaire is designed to have students select an appropriate answer to a question from a list that is provided, the questionnaire looks similar to a multiple-choice test that covers instructional effectiveness. An example of this type of questionnaire item is:

During this class, the instructor was able to keep my interest in the subject matter.

_____ all of the time

_____ most of the time

_____ some of the time

_____ none of the time

Questionnaires can also use open-ended questions. Examples of open-ended questions that are effective for evaluating instruction include:

- What are the major strengths of the instructor?
- What are the major weaknesses of the instructor?
- What do you suggest to improve this course?
- What aspects of the course were most valuable?
- What aspects of the course were least valuable?

Peer Assessment Of Instruction

Peer instructors are well-qualified to assess instruction. They have experiences dealing with students who have similar motivation, maturity, and ability levels as the instructor being evaluated. They also have instructional experience in planning and delivering instruction and in evaluating learning.

While peer instructors are very capable evaluators of instruction, they are seldom used as effectively as they could be. It is very difficult for an instructor to be "put on display" for a peer. An instructor's reputation as an instructor could be destroyed by the peer evaluator who does not understand the reason for evaluating instruction. When asking a peer to observe one's instruction, it is important that the purpose of the evaluation be made explicit. The peer evaluator must understand that the purpose of the evaluation is to improve instruction, what types of skills and competencies are to be evaluated, and that the observation is to be confidential.

Peer observations use a data-collection technique called systematic observation. This technique requires the observer to sit in on a class and observe the behavior of the instructor and the students. The information collected through a systematic observation is controlled through the use of either a checklist or a rating scale.

Checklists can be used effectively by peer evaluators when the focus of the evaluation is on the number of instructional behaviors that are displayed. Peer evaluators can use a checklist to record the instructional characteristics as they occur during personal observations. The peer evaluator would, for example, use the checklist to record the number of questions asked of students, the number of different students helped during class, and the amount of time spent on actual instruction. For example, a typical set of behaviors that might be included on such an evaluation instrument include directiveness, openness, enthusiasm, and flexibility. The peer evaluator could also use the checklist to record the behaviors of the students during the class. This type of assessment can provide valuable information regarding instructional quality because it focuses on the attention and interest of the students. When checklists are used to record student behaviors the evaluator would be interested in the number of times students talk to each other regarding non-instructional topics, the amount of time actually spent on learning, and the number of questions asked by the students. Following the completion of the evaluation, the results can be tallied and discussed between the observer and the instructor. Decisions can then be made regarding the needed areas of change.

A second type of instrument used by peer evaluators is a rating scale. The major difference between checklists and rating scales is that checklists measure the frequency of behaviors while rating scales measure the quality of the behavior. Rating scales are useful for evaluating behaviors that are not easily divided into specific items. For example, an observer may be asked to make a judgment on the degree of enthusiasm displayed by the instructor toward the subject matter. An evaluation of this type involves collection of subjective data. Because subjective data is neither inherently valid nor reliable, considerable care must be taken when developing rating scales.

Rating scales consist of a set of questions or statements that guide the observer during the evaluation. For each of the questions or statements, the observer must make a judgment based on knowledge and experience and then rate the instructor on the scale. Scales are typically constructed by using either a set of words to describe the attribute being evaluated or a set of numbers that relate to the quality of the behavior. Examples of rating scales that use words and number scales to describe an instructor's behavior is shown below:

The instructor's ability to provide clear, accurate, and relevant examples was:

outstanding   above average   average   below average   unsatisfactory

The degree of preparedness displayed by the instructor during the lesson was:

5	4	3	2	1
very well prepared			very poorly prepared	

Self-Assessment Through Videotaping and Reflection

The person who is closest to the actual instruction is, of course, the instructor. Self-assessment can be very effective for identifying instructional areas that need improvement. Self-assessment is also the easiest and least obtrusive type of instructional evaluation. Good instructors always reflect on how a particular class session went, how much students enjoyed the lesson, and how much they actually learned. While instructors are constantly evaluating their own instruction, they often do not use self-assessments as effectively as they could. Self-assessments are typically made on the spur of the moment and on a random basis. The information collected during these unplanned assessments vary greatly in type and quality. Even when the assessment information is obtained, instructors do not effectively nor efficiently use the information. Often they will obtain very valuable information while reflecting on the effectiveness of a particular instructional method yet will not use the information to make improvements.

Even with these apparent drawbacks, the individual instructor is one of the most qualified persons to assess his/her own instruction. The instructor does the planning, makes the actual delivery, observes and interacts with the students, and obtains firsthand information about the effectiveness of the instruction through the assessment of learning. In order to be effective as a self-assessor, the instructor must consciously plan the assessment and must take the time needed to analyze the evaluation information and make the necessary changes in instruction.

One of the best ways to self-assess instruction is to video tape a lesson. This is a simple technique that requires setting a video camera in the back of the classroom so it will show both the instructor and the students. If the camera has a wide-angle lens, it is not even necessary to have someone operate the camera during the instruction. The camera can be turned on at the beginning of the lesson and can be left on until the lesson is completed.

Following the recording of the lesson, the instructor can review the tape to observe strengths and weaknesses. Questions to ask when watching the tape include:

• Did you talk to the class or to the floor?

• Did you speak clearly and use your voice effectively?

- Was the information you wrote on the board legible from the back of the room?

- Did you recognize and respond to students' comments and questions?

- Do you appear to come across to students the way you thought you did?

Videotaping is a convenient, inexpensive, and accurate way to gather data on all of these questions.

Utilization Of The Results

When the assessment data has been collected, the data must be analyzed and interpreted. The analysis of the assessment results can be done either individually or in groups. For self-assessment data, it is most appropriate to review the data individually. For peer-assessment data, joint analysis and interpretation is most beneficial.

The major purpose of formative assessment of instruction is to identify weaknesses in instruction so improvements can be made. The data must be thoroughly examined to identify the weakest areas of instruction. Any negative comments and behaviors that are identified on the checklists, rating scales, and questionnaires must be considered as an area that might need improvement. During the identification of the weaknesses, the instructor must keep an open mind. The weaknesses that will be identified should be viewed as constructive criticism rather than as personal attacks. It must be remembered that improvements in instruction cannot be made until the weaknesses are identified. Once the weaknesses have been identified, the instructor can begin to address those areas through changes in the planning, delivery, or assessment of instruction.

REACHING THE GOAL OF HIGH QUALITY ASSESSMENT

Much of the assessment that occurs in construction technology courses closely resembles authentic assessment. Technology teachers give assignments to students that are imprecise, problem-centered, and highly experiential. Students complete these assignments with considerable problem solving and other intellectual-process activity as well as hands-on practice with tools, materials, and equipment. During the completion of these activities, technology teachers observe students in action, provide mentoring and coaching support, and continually assess student performance. Much of the subjective, naturalistic assessment that occurs through observations and interactions with students in laboratory settings provides technology teachers with a greater understanding of student learning that can be used to improve instruction and to assign grades. While the assessment methods used in construction technology courses is quite good, alternative and authentic assessment techniques can make it better. Technology teachers should strive to continually improve their ability to assess learning and instruction.

REFERENCES

Collins, A. (1990). Reformulating testing to measure learning and thinking. In N. Frederiksen, R. Glaser, A. Lesgold, & M.G. Shafto (Eds.), *Diagnostic monitoring of skill and knowledge acquisition,* (pp. 75–87). Hillsdale, NJ: Erlbaum.

Erickson, R.C., & Wentling, T.L. (1988). *Measuring student growth: Techniques and procedures for occupational education.* Urbana, IL: Griffon Press.

Frederiksen, J.R., & White, B.Y. (1990). Intelligent tutors as intelligent testers. In N. Frederiksen, R. Glaser, A. Lesgold, & M.G. Shafto (Eds.), *Diagnostic monitoring of skill and knowledge acquisition,* (pp. 1–25). Hillsdale, NJ: Erlbaum.

Frederiksen, N. (1990). Introduction. In N. Frederiksen, R. Glaser, A. Lesgold, & M.G. Shafto (Eds.), *Diagnostic monitoring of skill and knowledge acquisition,* (pp. ix-xvii). Hillsdale, NJ: Erlbaum.

Henak, R.M. (1993). *Exploring construction.* South Holland, IL: Goodheart-Willcox.

Kirst, M.W. (1991). Interview on assessment with Lorrie Shepard. *Educational Researcher, 20*(2), 21–23, 27.

Linn, R.L. (1991). Dimensions of thinking: Implications for testing. In L. Idol & B.F. Jones (Eds.), *Dimensions of thinking and cognitive instruction,* (pp. 179–208). Hillsdale, NJ: Erlbaum.

McGaghie, W.C. (1991). Professional competence evaluation. *Educational Researcher, 20*(1), 3–9.

Stenmark, J.K. (1989). *Assessment alternatives in mathematics.* Berkeley: University of California.

Whitehead, A.N. (1929). *The aims of education.* New York: Macmillan.

Wolf, D., Bixby, J., Glenn, J., & Gardner, H. (1991). To use their minds well: Investigating new forms of student assessment. In G. Grant (Ed.), *Review of research in education,* Vol. 17 (pp. 31–74). Washington, DC: American Educational Research Association.

11

Construction Technology In Developing Countries

Alfredo R. Missair, M.Arch.
Department of Architecture
Ball State University, Muncie, IN

In all latitudes of our planet, construction has been a basic activity of humankind and is intricately linked with the necessity for shelter which assures day-to-day survival. We know that the act of constructing and building structures has evolved under all types of climatic conditions. Likewise, builders using much of the earth's available resources have produced the most obvious material legacy—a tangible testimony of all cultures throughout history. In many regions of the world, and before the written word was developed native builders adopted western civilization as their ideal or had it imposed upon them as the model. Construction patterns, techniques, and technologies were developed and transmitted from one generation to the next. Therefore, the activity of constructing with the development and transfer of technology can be understood and analyzed as a phenomenon. Similar to a language, this activity is naturally an integral part of the all-encompassing phenomenon of cultural development.

A HISTORICAL OVERVIEW

Just as the developed-world countries did in earlier times, the cultures of the undeveloped world, created a complex and sophisticated relationship between the environment, resources, technology, construction, education, and culture through generations and patient experimentation. One of the key concepts found in these relationships was the idea of sustainability. Sustainability was more the result of the unconscious and the self-centered response of people to the basic need for survival, than a goal in itself. This search for a permanent equilibrium propelled various groups of humans in their quest to provide themselves with a safe and predictable environment to grow and prosper.

Unfortunately, these unique technologies used and developed by early vernacular builders were sadly misunderstood and considered primitive by the early western explorers and conquistadors. These techniques were dismissed as inferior and relegated to oblivion. Western explorers were unable to understand the sophisticated relationship that existed between all the tangible and intangible elements, and the processes involved in the construction activities of other cultures. By definition, these individuals were not part of the cultural and generational educational process of technological transfer to the new territories. The following examples will certainly illustrate the complex relationships which existed between nature, humans, and man-made settings in ancient societies of the developing world.

Example One: The Mayas

The Mayas (1,000 B.C. to 1,600 A.D.) resided in the city of Tical, which is today the Republic of Guatemala. They achieved an unparalleled understanding of a sustainable and productive semi-urban agricultural ecosystem. This unique approach included the mastery of construction technology for raised fields, or chinampas, which were surrounded by a grid of water channels used for growing agricultural produce and raising fish for large-scale human consumption. The Mayas' sustainable technology was based on the fundamental understanding of the local biological equilibrium which existed between the water lily and the mojarra (a small protein-rich fish) that fed on the roots of the water lily. The water lilies grew naturally around the perimeter of the small island-like family-operated farm and were used to produce compost. Furthermore, the mojarras provided the protein that complemented the basic cereal and vegetable diet obtained from the crops grown in these raised fields. So important were both the water lilies and the mojarras, that they became the symbols of power of Tical's ancient carved reliefs depicting the princes who ruled the Mayas. This example illustrates the delicate relationship that existed between the construction processes of the raised fields and the sustainable food source of the mojarras.

Example Two: The Tiahuanacos

In Bolivia, the Tiahuanacos (600 A.D.), whose sophisticated civilization preceded the rise of the Inca Empire, also mastered a construction technology that sustained productive farmland. The civilization was located at 4,200 meters of altitude in the meseta Andina (high plateau), in the area which surrounds Lake Titicaca. Like the Mayas, the Tiahuanacos invented a construction technology of raised fields parallel to single yet large water channels. The raised fields and water-channel system created a reliable ecosystem that regulated the abrupt changes of daily temperature in the semi-arid climatic condition of the high Andes. These ancient builders carefully controlled layers of sand, clay, and aggregates in order to build a gigantic reverse percolating system that supplied water to the roots of the crops grown above the water level of the channel.

It is also interesting to note, that the above mentioned environmental approach to construction technology was developed centuries before any western civilized person had discovered the New World. Similarly, the use of contemporary technological concepts such as composite materials are presently employed in new western industrial developments. These are found in many mud walls, built with vernacular construction techniques, around the world. Some of the complex ancient construction systems are still in use on the African continent and remain superior to anything currently used in the developed world.

Example Three: The Swahili

A third example is the biodegradable urban sanitary systems located in east Africa. These systems biologically controlled endemic diseases long before any similar idea was conceptualized in Europe. At this point, a detailed description of this unique man-made urban environment is necessary if we are to understand the belief that man's construction technology can harmoniously sustain development when the environmental and cultural conditions are respected in any given region of the world.

Along the Kenyan coast of the Indian Ocean, a unique culture, called the Swahili, developed after the ninth century A.D. This culture was the result of the racial mixing of mainly the Bantus and Arabs, and embraced Islam as their religious and cultural identity. Certainly one of the most remarkable achievements of this culture was to be produced through its architectural and passive construction technologies. They flourished inside the famous city-states of the coastal islands such as Zanzibar and Pemba, and particularly in settlements of Kenya in the Lamu Archipelago along the northern coast.

The town of Lamu, located on the island of the same name, has remained curiously unchanged throughout passage of time. It was indifferent to western technical progress and modernization. The town's unaltered dense urban structure and architectural forms were defined by extremely narrow streets which prohibited the introduction of motor vehicles to the island. To some extent, Lamu survived while ignoring the industrial revolution. Although the earliest buildings of Lamu have disappeared, most of the coral houses dating from the 18th and 19th centuries are still inhabited. These structures delineate the characteristic urban fabric of this historical town. One of the very special features of these houses is their sophisticated bathrooms and their unique sanitary system. They have contributed significantly to the long-term survival of a densely populated urban society with limited technological resources.

The design of the houses is strongly based on the privacy gradient, a feature typical of most Islamic societies. In their use of this design principle, Lamu people remain very devout religious adherents. Externally, a massive facade, sometimes interrupted high above the street level by small openings, is complemented internally by a layout based on the privacy gradient which protects the houses from public observation.

Adjacent houses share side walls giving continuity and homogeneity to the street front. In adherence to the strict Islamic teachings concerning the control of privacy, all of the layers of family social life are organized in the house plan. The space most open to the public at the street level is called "daka." The most private room is called "ndani," which houses the women of the harem. The daka is the access or open-external vestibule that signals the entrance to the house without disturbing the peaceful ambience of the interior of the dwelling. This typical feature of the Swahili architecture was designed as a bold negative space, and is nicely accentuated by one or two masonry benches called "barazas." The space is also accented with a massive double-wooden door with a finely carved frame and central post. The daka is also the place that is designed to enable the dwelling's inhabitants to exert gradual control over the public and social interaction with the community. During much of the evening, relaxed conversations occur in these spaces. They also serve to punctuate the narrow streets as harbors for social life. Located immediately beyond the door is a room called "tekani" which serves a dual purpose. It is a buffer space for the reception of the family's male friends, and shelters the women from sight. Quite often, the first bathroom of the house and a guest room, the "sebule," will be adjacent and accessible to that internal vestibule. Thus, all necessary comforts are provided for the visitor while the life of the family's women is kept undisturbed. All direct views from the street and entrance are prevented, and only the increasing brightness preceded by a cool sensation will suggest the patio beyond. This patio is named the "kiwanda." The kiwanda is the central open space that defines and articulates the more public spaces of the front of the house. Therefore, it serves to set the territorial boundary of the increasingly private family quarters at the back of the house. Light, ventilation, and controlled environmental conditions are provided which make it possible to grow vegetation and cool the dwelling.

Opposite the public area and across the kiwanda, are a series of elongated chambers connected centrally by large aligned openings. On each side of the most distant and more private chambers is a bed space which is usually protected by a loose curtain hanging from a cylindrical pole. These spaces are the equivalent of bedrooms in western nations. The number of chambers can vary from house to house. The one closest to the patio, called the "msana wa tini," is followed by the "msana wa yuu." Finally, the most elaborate of all, the "ndani," is reserved for the women of the family. At the end a sculptured wall, the "zidaka," embellishes the spatial sequence throughout the openings displaying rows of carved small niches. A sophisticated, excellent example of plaster work and geometry is used to create false perspective effects that compensate for the narrow spatial alignment of walls. Women use the zidaka to display their fine china dishes. It also serves as the inspiration for pieces of traditional Kiswahili poetry (Ghaidan, 1976, p. 5).

The natural characteristics of the materials available in the archipelago are formed by coral reefs, and have dictated the architectural form of the typical Swahili house. Specially grown mangroves provide straight poles called "boriti," which serve as horizontal structural components supporting stratified coral slabs topped with a mixture of coral rag and lime mortar. These slabs form the upper floor and the flat roof. Because of the boriti's natural strength and the heavy load that it must support, the dimensions of the interior spaces are significantly limited and, typically, span no more than 2.5 to 3.0 meters. This restriction causes a formation of a constant modular dimension perpendicular to the street front. In the other axis, the spacing between each boriti member is usually 0.25 to 0.30 meters and is repeated, as many times as needed to cover the total length of the urban plot frontage. In order to support these horizontal structures, heavy walls are constructed with coral rag and mortar in thickness close to a cubit (0.50 meters). This curious functional module and proportional system, at plan level of 1:3 to 1:4 ratio, emerged from these structural limitations. The spatial sequencing of the domestic spaces is planned so these elongated rectangular modules can be connected along a strong transverse axis. This results in the creation of interesting layering effects, and as noted before, strongly flattened perspectives. J. de Vere Allen (1978), ex-curator of the Lamu Museum, pointed out that,

> Undoubtedly the most impressive part of the houses is their bathroom and lavatory system. The bathrooms were especially beautiful, with plaster work, niches, and sometimes even carved coral decorations. The money spent on them can only have been equalled by the money

spent on quiblas (sacred niche in wall oriented to Mecca) of Mosques that, oddly enough, with their trifoliate arches, their chamfered pilasters . . . they often resemble. The link between cleanliness and Godliness, [being] always strong in Islam. (p. 12)

Not only were the bathrooms of the coral historic houses abundant (two to four per residence), but they were also creatively designed. They are a successful and unique example of pit latrines constructed inside the dwelling itself. Surprisingly, many of the bathroom chambers, with their attached latrine were located in the upper level of the houses, were connected through a vertical shaft to the ground-level pits. This vertical layout demanded careful planning to resolve the functional organization and sanitation system.

The latrines are systematically placed at the periphery of the house along the external coral-rag walls. According to a technical study conducted by the Housing Research and Development Unit of the University of Nairobi (1980) entitled, *The pit latrines of Lamu: 600 years of 'illegal' sanitation,* "the layout was planned from inception to allow for the excavation of the temporary outdoor shaft, which would momentarily block a narrow street or an open patio to enable the easy emptying of the existing pits when filled" (p. 9). It is also believed that part of the content of the pits is recycled naturally through a combined action of the ocean tides and underground sandy layers. Furthermore, the same study, states that, "another remarkable observation is the infrequency with which many of these pits require emptying, . . . figures of up to 45 years were given" (p. 11).

The access to the bathrooms usually takes place through a rectangular opening at a coral-rag wall close to the entrance of the house or at the extremity of the ndani. Often, a black, white, and red painted square-shaped wooden lintel is displayed as an elegant, but brief, statement of woodworking excellence. Traces of cylindrical holes can be found in some bathrooms at the door-sill and lintel and are believed to have held the wooden hinges of the bathroom doors which are presently missing.

Facing the entrance door, a "birika" or floor-level water cistern can be seen along the major wall of the main bathroom chamber. Clear water is kept fresh, free of mosquito larvae, thanks to the presence of a special breed of very small fish inside the birika. This utilization of small living creatures to destroy the vector that spreads malaria can be considered one of the earliest systematic and natural preventive approaches used to control an endemic disease. So important was the survival of these fishes to the Swahili, that an imported china bowl would be placed at the bottom of the birikas to save the last available amount of water in which the small fish took refuge during periods of severe drought or when cleaning the cistern.

Architect and historian, Usam Ghaidan (1976), in his book, *Lamu: A Study in Conservation,* briefly describes the principle by which water was supplied to the birikas. The sweet water came from a nearby well or was collected from the roof and, then, channelled through a specially designed funnel and stone conduit to an internal cistern in each bathroom. A low-wall partition which was constructed at a height 1.20 to 1.80 meters to separate the birika from the latrine and provide privacy. Traditionally called "dado" in Kiswahili, its upper part was finely carved with regular geometric patterns of primarily square shapes forming a frieze that usually ended in a small arched niche (see Figure 11–1). The actual access to the latrine itself is very often one of the most elaborate and carefully designed components of the bathrooms. It can be seen either as a regular or a trifoliate arch supported

Figure 11–1:   Detail of dado wall dividing the entrance of the latrine.

on chamfered pillars. The half-dome shaped ceiling covering the alcove, which housed the squatting plate of the latrine, sometimes had its surface carved with regularly spaced horizontal strips and lines. One, two, or three small funnel shaped vertical openings found in either one or all of the three walls that formed the latrine provide ventilation and highly controlled light without sacrificing privacy. These openings also served—intentionally or unintentionally—to attract the light-seeking flies outside the somber latrines which aided in the control of the most common vector that spreads gastrointestinal diseases and is the major cause of morbidity in the tropics.

In order to guarantee personal hygiene and regular body functions, the amount of water necessitated for the efficient use of the bathroom is insured by a rational utilization of the natural supply available in the island. A large spoon-like container made from the shell of a coconut and attached to a wooden stick was usually found atop the front birika's parapet. This natural measuring device was sufficient to guarantee personal cleanliness and also provided the right amount of water for the anal cleansing practiced in the region.

Lamu sustained a continuously high density—estimated at 260 persons/hectare and a healthy population for more than 600 years by relying on those internal pit latrines, the water cisterns with the small fish, and the careful consumption of scarce water resources (England, 1980, p. 39). Figures 11–2 and 11–3 provide illustrations of the Lamu bathrooms.

These observations confirm the uniqueness of the bathroom's creative formal expression as well as its functional design which is complemented by the natural composition of layered sand and coral of the island, and the tidal cycle of the Indian Ocean. Integrating architecture, society, and environment, the Swahili worked in harmony with their ecosystem to produce an exceptionally beautiful and sophisticated technical and architectural solution for a commonly accepted bleak space in our western modern dwellings. It could be stated that the balance between the dense urban setting of Lamu and its subtle social fabric is due to a harmonious relationship between natural phenomena and the physical form within the framework of limited resources. These conditions challenged the local inventiveness as well as its intrinsic artistic expression to produce a unique approach to architectural design and construction technology.

Technology Transfer

Negative Effects.   More recently, the town of Lamu has become a tourist attraction. Some of the unquestioned modern western comforts and progress are being systematically introduced. Piped water is captured from a limited source in the island underneath sand dunes of Shela, a small town some

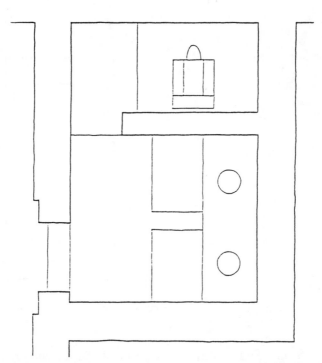

Figure 11–2:   Floor plan of eighteenth century Lamu coral bathroom.

miles away. There is no evidence today of the amount and permanence of the supply. Pedestal toilet fixtures are being installed on top of pit latrines inside historic coral houses bought by vacationers. Slowly, the traditional sanitation system is being transformed. The new system will function with great dependency on large amounts of water and will produce similar amounts of sewage. The efficiency of the internal pit latrines could be jeopardized due to the amount of water that is released every time the toilet fixtures are used. The individual reserves of water in the birikas seldom exist anymore. Because of their convenient size and location, these cisterns are systematically transformed into bathtubs that also usually discharge their contents of soiled water and soap into the pits of the adjacent latrines (England, 1980).

A unique environmentally sound construction technology with its particular architectural expression is going to be irreversibly transformed toward a technology dependent on a far more contaminating substitute.

Figure 11–3: Sample elevations of trifoliate arch leading to the latrine alcove.

Positive Effects. During July and August of 1990, this author served as a consultant for the United Nations in Bangladesh. The purpose of the mission was to review a low-income human settlement project that included as many as 2,600 double pit latrines located outside the core-houses built for an Islamic community. It is believed, that if this information had been available at that time, recommendations would have included careful consideration of more creative approaches to the design and engineering of bathrooms and sanitation systems which were based on the harmonious and sustainable relationship that exists between nature, culture, and resources as exemplified in the Swahili coral houses.

THE EUROPEAN COLONIAL INFLUENCE

The discovery of the West Indies or New World, and the systematic exploration, conquest, and later exploitation of Africa and Asia was accomplished with a single underlying goal—the rapid seizure and possession of the material wealth as it was known and pre-defined by the European powers. The basic purpose of this exploration was not to access new knowledge that could be transferred to the Old World, but rather it was the shipment of material wealth to Europe in the form of tomatoes, potatoes, cacao, silver, gold, and precious stones. Simultaneously, a new era of systematic European scientific progress based on the development of cartography, astronomy, and navigation necessary to achieve these explorations may have also contributed to the disregard of the subtle local technologies that had been previously unknown to the Old World.

Today, no one can assess what could have been the impact of ancient interdisciplinary approaches to a sustainable development based in ecologically fit construction technologies. It is certain that, if the European scholars could have noticed or seen the intangible relationships which existed between nature and technology as experienced by other cultures, we would have a much different world today. But, as the ancient African proverb says: "The foreigner only sees what he already knows," and the early European explorers were no exception. Worse yet, much of the information that could have been collected from local documents, such as the thousands of Aztec's books of ideograms, were systematically burned in the name of the new religious order and moral values imposed upon the conquered people.

Some construction technologies useful to the major European powers of the "age of the discoverers" were introduced in the conquered territories. In some early cases, the transferred construction technology had a notorious latent capacity to be adapted to the new environments. A typical example is the Spanish introduction of the fired-clay brick and roof tile in Latin America. Today in South America, these materials are considered the typical choice for local construction.

In some countries like Colombia and Uruguay, local contemporary architects and builders excel in the mastery and development of new technologies based in the fired-clay brick. At the forefront is the architect/engineer Eladio Dieste from Uruguay, who has pushed the structural strength and aesthetical quality of the brick masonry to unexpected limits. Famous are the Dieste structures characterized by big spans, membrane-like structural wave-shaped walls, and sophisticated natural-light treatments of industrial and religious buildings.

The history of the development of brick masonry in the Latin American continent is a successful story of technological transfer across the Atlantic Ocean that remains unmatched in terms of utilization of local resources and minimal consumption of non-renewable energy for its fabrication. The tremendous success of the fired-clay brick was predictable from its original conception. The only thing necessary to transfer was the knowledge of the production process and the constructive geometry to apply the finished product. At times of rapid but arduous territorial expansion, it was not necessary to wait for the slow and unpredictable arrival of sailing ships carrying building materials from Spain or Portugal because the principal components for brick fabrication — clay, water, fire, and manpower were available locally. Availability of necessary materials and labor coupled with Spanish town planning resulted in the successful urbanization phenomenon of the New World.

The Spanish town-planning strategy was based upon the two-dimensional planning grid or "gridiron" with individual urban plots of 10 "varas" (each vara or stick measured around 86 cm), and the three-dimensional constructive module of the brick. Today the physical impact of these two basic modules the brick and the vara, are perceptible in the cityscapes and built-forms of Latin America.

During the industrial revolution, the concept and form of the European influence on developing regions and colonies of the world were reversed from the issue of using locally available resources. In fact, the main objective was a total substitution of the local know-how and resources for western technologies and materials. The late colonial power's approach to their vision of world development was that local raw materials had to be exported to the Old World markets and manufactured products would be imported to their overseas possessions.

A good example of this policy was the fantastic flow of corrugated metal sheets from Europe to Africa during the industrial era. Builders readily adopted the light-weight thin metallic membrane because it added significant structural integrity. Also, the geometric applications of the metal sheets allowed the builder a maximum amount of sheets for minimum space occupancy and, therefore minimum freight cost. In short, it was an ideal material to export.

Although the metallic corrugated sheet technology disregarded the principles of thermal insulation in the tropics, these sheets became the material of choice and replaced the primitive thatch roofs of vernacular constructions. This trend continues strongly today. With ease of replacement, maintenance, and low initial cost, it is being chosen in spite of resultant unbearable thermal conditions. In the least developed countries, the use of corrugated steel sheets still represents the first step toward the very publicized conveniences of western construction technologies.

In the early years of the industrial revolution when electro-mechanical inventions were pioneered, it was difficult to transfer western technological developments to the inhospitable regions of the developing world, due to the lack of infrastructure support. Instead, builders used common sense and passive construction designs in these regions in lieu of the many luxuries of the new industrial era. An analysis of buildings constructed by French expatriates in West Africa is a lesson in employment of simple but effective design concepts and construction techniques that have assured the maximum comfort with the minimal dependency upon electro-mechanical technology. Obvious strategies included the use of large verandas at the periphery of the buildings. The removable shading devices that allowed ventilation and control of direct light, the avoidance of internal circulations, and thick masonry walls for thermal storage of the cool night temperatures.

The general education and technical training available to the colonies was usually accessible only by the privileged few and required long stays in the capitals of Europe. Future technicians, who would be going to work in their native lands, where economic conditions and environments were totally different were instructed in the latest construction technologies of the western nations.

ACCESS TO NATIONHOOD AND DEPENDENT "INDEPENDENCE"

After winning political independence, the new nations had to confront a new type of colonialism calculated to make fast profits for the west but minimal benefits for the developing country. Examples are the famous white elephants or turn-key projects built in Africa. Large hospitals requiring complex technologies, imported materials, and gigantic but non-existent annual budgets for maintenance and recurrent costs were designed for the poorest countries. Facilities, with no physicians or nurses to run them, were built where no sewage existed. Toilets oriented to Mecca were designed for Muslim people and built in ignorance of the local culture and religious beliefs. Concrete blocks became the standard building material in Africa, south of the Sahara, where only one cement factory existed. The former Soviet Union even included snow plows in turn-key factories built in equatorial Africa. The U.S. exported non-metric building materials. When a user of the building would lose the key of a door lock, he or she would have to buy a new door. The European construction industry insisted upon its high-technology construction standards favoring implicitly the import of European-made construction components.

Things did not change much for the young nations of Africa and Asia in the late 50's and early 60's. In order to achieve reasonable levels of comfort in their western technology-dependent building projects, future technicians were systematically educated to design and build in cold-temperate climates using as much energy as needed. This energy was dependent upon non-renewable imported resources. The adoption of high-technology systems, similar to those commonly used in the west for construction activities and the building industry created the problem of non-sustainable maintenance. Sights of decrepit installations and buildings became familiar. It became necessary for foreign technicians from Europe to be employed for building of all large-sized construction projects and to manage any technical facility including hospitals, factories, and power plants. Preventive maintenance procedures and budgets were seldom a concern during project development. If we were to consider that a general hospital facility of moderate size would need to cover its recurrent costs and a budgeted annual expenditure close to one-third of the initial capital investment for the facility's construction and equipment, we can easily figure the devastating impact that the transfer of non-appropriate and non-sustainable technologies had upon the treasuries of the newly independent nations. This post-World War II trend of complete substitution of people, resources, and know-how compromised the real access to independence for many young nations of the developing world.

Another damaging issue for developing countries has been the approach to the construction financing of major public-housing schemes. Many of these were built using foreign standards and materials; and disregarding local needs, resources, cultural and social patterns, and economic and financial possibilities. Only adaptations to the climatic conditions were implemented in some well-intended cases. The paradox is that a common measure of a nation's economic health is usually assessed in terms of productivity of its housing and construction industry. Substitution of locally produced houses with imported models and the use of foreign currency to build them have been reasons for the current state of affairs in some developing nations. Urban plans, architectural projects, and construction technologies in these countries, as in any other country, must first target the growth and development of their communities using their resources and employing their people, in order to generate stable sources of income and socio-economic progress.

THE RECOGNITION OF LOCAL VALUES

Young nations have had to confront two major roadblocks in their attempt to access development of appropriate technologies. The first roadblock was the need to rediscover and relearn the techniques and fundamental rationale inherent in their traditional construction technologies. The second roadblock was to understand that the physical heritage in the form of buildings and construction technology know-how of the Colonial era, was now part of their historical national evolution. Colonial-era buildings, as mentioned before, sometimes represent sensitive and non-dependent solutions for difficult construction problems.

For example, some years ago this author had an experience which exemplified this type of solution. A young African scholar, serving as a government housing official, explained to me the appropriateness of a floor plan produced for early French colonial buildings in the Republic of Mali, in West Africa. That embarrassing legacy was a curious incongruity. After years of having been ignored by the newly independent generations, because they had housed institutions and nationals of the oppressing Colonial powers, they were now a permanent part of their architectural heritage.

A significant contribution to the recovery and revalidation process occurred in Egypt, thanks to the work of the architect-scholar Hassan Fathy (1973), who published his now famous work entitled *Architecture for the Poor.* Fathy uncovered the ancient mud-brick construction technology known as the theory of the lining arches, which was used in ancient Egypt more than 3,000 years ago. This construction technology used the least expensive construction material in the world — earth, which made possible the erection of extraordinary buildings from gigantic covered water cisterns to medium-density housing. The lining-arches theory is based upon the mastery and knowledge of spherical three-dimensional construction geometry. Parallel to this development, scholars and technicians from both the developing nations and the first-world study appropriate technologies based on a sustainable approach to growth and development within the larger realm of cultural-fit. Work has been produced from the inclusion of fiber from the coconut shell in industrialized panels to the use of bamboo as a substitute for iron rods in the construction of concrete slabs or the addition of enzymes in the production of mud-bricks to make them moisture-proof.

THE PATH TOWARD A SUSTAINABLE DEVELOPMENT

Today, the knowledge and understanding of local, natural products and low-technology processes are determinants in the developing nation's success in gaining access to a sustainable development. Many products of the advanced technology industries are used as vehicles for more effective implementation of sustainable development projects. For example, much is to be gained through the employment of passive-solar systems to provide energy to nations without oil resources. Since the development of photo-voltaic solar panels, the most significant civilian use of this reliable technology has been in developing nations such as the Gambia Republic in West Africa. Photo-voltaic solar panels in that country provide a sustainable electrical energy supply for portable vaccine refrigerators, fluorescent lights in emergency delivery quarters, and telephones for ambulance dispatch requests. Mass inoculation of children with fresh and reliable vaccines throughout the country was finally made possible because the unreliable kerosene refrigerators were replaced. Infant mortality has declined over a 15 year period from close to 250 per thousand to less than 100 per thousand since the adoption of total solarization of the dispensaries and cold chain in the country, and the introduction of better health-care habits (Medical Research Council, Infant Mortality Rate, 1989). Other growing uses of solar technology in the developing world near the Equator include solar-water pumps and solar-water purification systems. Another significant technological contribution being employed to remedy the financial paralysis that precluded development in third-world cities is the recent implementation of computer-based cadastres and data banks using geographical information systems and computer-aided-design software. This approach facilitates the processing of information and better management of already scarce resources for city growth and improvement by combining socio-economic data with services and infrastructure data; "smart maps" are being produced.

SUMMARY

Along with the recent consolidation of democratic processes in the Third World, there is a conscious understanding that substitutive transfers of technology and dependency-based help are outmoded policies. The obvious failures of a multitude of western-conceived or inspired development projects has caused many of the parties involved — the donor countries or agencies, their counterparts or local governments, or the supposed beneficiaries — to develop a suspicious or negative attitude. Measurement of success in development projects is also an extremely difficult task, one that should consider the evolutionary nature of human betterment. Human development can only be understood in a long-term approach that considers generational changes that do not necessarily follow linear paths. Measurable improvements to reassure the donor or lending countries, international institutions, and the industrialized nations should use indicators based on valid criteria and formulated with the assistance of the targeted beneficiaries. Naturally, these criteria are not always quantifiable and should be understood within the broader vision of the evolutionary context of human development.

Thus, cultural considerations for understanding the intangible processes inherent in the utilization of any technology that may be required for human development are a necessary basis for the achievement of a sustainable development. Development will follow the evolutionary nature of culture and progress if consideration is given to those who will benefit. This linkage between technology, culture, and direct beneficiaries seems in retrospect to be the missing link in most of the development projects that have failed in the past or, even worse, have produced long-term cultural, environmental, or economic damage.

Today, the rationale of multilateral international organizations for the formulation of development projects is aimed toward the utilization of local resources and the work force. Culturally appropriated technical assistance and technologies are now the main vehicles used to attain project replicability and sustainability. The outcome of well-conceived projects in developing nations should achieve a positive impact for the long-term socioeconomic development of the beneficiaries based on the sensible respect for the cultures in the communities concerned. South-to-south collaboration and training, accessibility to employment, personal credit, continuing education, community participation and development, consideration of maintenance costs, economic and financial replicability, and respect for the culture are all necessary seeds for successful project implementation resulting in an equitable, sustainable development.

REFERENCES

Boorstin, D.J. (1983). *The discoverers.* New York: Random House.

Caminos, H., & Goethert R. (1978). *Urbanization primer.* Cambridge: The MIT Press.

de Vere Allen, J. (1978). *Lamu town: A guide.* Nairobi: author.

Doutrewe, S., & FranÁoise, R. (1985). *Architecture coloniale en cute d'ivoire.* Abidjan: Les Publications du MinistÉre des Affaires Culturelles.

England, R. et al. (1980). *The pit latrines of Lamu: 600 years of 'illegal' sanitation.* Nairobi: HRDU, University of Nairobi.

Fathy, H. (1973). *Architecture for the poor.* Chicago, IL: University of Chicago Press.

Fedders, A., & Salvadori, C. (1980). *Peoples and cultures of Kenya.* Nairobi: Transafrica.

Ghaidan, U. (1976). *Lamu: A study in conservation.* Nairobi: East African Literature Bureau.

Kleczkowski, B.M., & Pibouleau, R. (Eds.). (1979). *Approaches to planning and design of health care facilities in developing areas.* (Volume 3) Geneva: World Health Organization.

Mayas, L. (1981). *Lords of the jungle.* New York: Odyssey Series.

Medical Reserach Council. (1989). *Infant mortality rate.* Geneva: World Health Organization.

Missair, A., & Bidwell, E. (1991). *Evaluation report: Primary health care facilities and solar-powered equipment in rural areas.* New York: United Nations.

Moseley, M. (1992). *The Incas and their ancestors: The archeology of Peru.* London: Thames and Hudson.

UNDP/HABITAT. (1990). *Field report: Una experiencia para orientar el desarrollo urbano y municipal.* Bolivia: Author.

Vetter, W.F. (1979). Advanced building techniques and their utilization in developing countries. In B.M. Kleczkowski & R. Pibouleau (Eds.), *Approaches to planning and design of health care facilities in developing areas* (pp. 115–117). Geneva: World Health Organization.

CHAPTER 12

Summary And Reflections

M. James Bensen, Ed.D.
President
Dunwoody Institute, Minneapolis, MN

This yearbook has provided a strong argument for the study of construction in technology education. Simply stated, mankind continues to use technology to construct structures on a site for a variety of purposes such as shelter, communication, transportation, storage, and recreation. This review of the major issues related to construction in technology education has been organized into seven major headings, which are as follows: rationale, development of construction in civilization, impacts of construction, construction in technology education, innovations in management, technological changes, and careers in construction.

Rationale

A rationale is the reason for being. It is the foundation upon which the curriculum is built. The rationale answers the why question. In the first chapter of this yearbook, Richard Henak outlines an excellent case for why construction needs to be taught in our schools. As we reflect upon why construction needs to be taught as a part of being technologically literate, the following perspectives were expanded.

Construction And The Development Of Civilization

A brief evolution of construction was described in the second chapter. As it does today, early construction took on the compelling needs of society. In order to survive, humans had to use adaptive systems to control conditions that surrounded them. The early cave dweller was a constructor when coverings were placed at the mouth of the cave to contain heat or provide protection. People in hunting and gathering societies were constructors when they built portable shelters which they dismantled and took with them so they could follow their food supply or move seasonally to more hospitable environments.

During the evolution from a society based on hunting and gathering to an agrarian society and later to an industrial society, rapid advances took place in construction. Structures were designed to last longer. The motivation moved from just providing shelter to that of doing business (e.g., craft and retail shops); easing movement and communication (e.g., roads, ferries, bridges, signal towers); controlling the environment (e.g., dikes, dams); providing places of worship (e.g., altars, churches, synagogues, cathedrals); and memorials (e.g., shrines, pyramids).

Today, we continue to make significant improvements in materials and expand our ideas on how to construct structures. At this point, construction technology appears to the lay person as wonders-to-behold as they observe structures taking shape. We tunnel through mountains or under water from two different directions and meet at a precise position; we put up structures that flex with the environment, and others that counter forces of wind or earthquakes with stabilizers. Daily, constructors continue to amaze society with the application of new technology.

The Impacts Of Construction

In the third chapter, Peter Wright expertly builds the case of the impact that construction has on our society. One of the interesting paradoxes of construction is that it is impacted by the economy and it, in turn, impacts the economy. As a value-added activity like manufacturing that contributes to the wealth of a society, construction is the type of value-adding activity that is often postponed when the economy is soft. Hence, construction has much more of an "ebb and flow" than manufacturing in impacting upon our nation's wealth.

The Environment. The environment in which we live encapsulates all human endeavors. Everything that we experience is shaped, tempered, expanded, and controlled by our environment. A review of the typical college-prep curriculum which tends to dominate what is deemed important in most schools, reveals that an assumption has been made that our lives are primarily impacted by the natural environment. Hence, we see years of science being taught to describe the laws and theorems that describe the natural environment. While the importance of the natural environment is not contested, what parallel do we see regarding the built environment?

The built environment requires an understanding of technology as it describes the know-how and rules of efficient action so that it can be understood by the learner. We realize that the built environment is subsumed by the natural environment, yet we also observe that we spend over 95% of our waking and sleeping hours in the built environment. As citizens of this shrinking world, we spend all this time in the built environment, and yet have a steadily fading understanding of it.

Value Adding. The ability for a society to generate wealth, increase its standard of living, and improve its lifestyle requires the ability to add value to its resources. To add value requires technology or what some refer to as the rules of efficient action. Countries with rich resources that can be extracted or harvested find that these resources do contribute initially to its wealth. But it is the know-how to conduct value-added processes to our resources that provides the most significant gains of wealth to society. This process is called production; and we produce the two elements of goods and services. It is the goods producing element of production that generates the significant wealth for a society through its value-adding activity. We do this either through manufacturing (producing in-plant) or construction (producing on-site). While the production of services is important in construction, particularly in maintaining and updating the structure and its systems, the initial construction act is the element that generates the first and greatest level of wealth. At this point, most people discover that manufacturing is viewed as more of a mystery than construction, because it takes place behind closed doors. Many times it involves working with sophisticated techniques such as lasers, robots, and computers. Though construction takes place out in the open for all to observe, it too, is becoming very sophisticated, is increasingly complex, and a growing number of its processes and techniques are not discernable to the "sidewalk superintendent's" eye.

Construction In Technology Education

Introducing construction technology into the education program of the pre-school-grade 12 curriculum can be complex. It is generally recommended that construction technology be integrated into the pre-school-grade 6 experience to support the subject areas of math, science, language arts and social studies. This is because the current curriculum is already so full that to ask for additional time to focus on technology education to any extent reaches resistance on the part of schools. However, if teachers can see that their students are learning faster and are able to handle higher-order thinking skills because of the experiences that they would get through technology, they are generally willing to use this powerful teaching tool in their instruction.

At the middle or junior high school level, special courses should be offered that feature construction along with other technology experiences. In the senior high school, the curriculum can take on more specialized functions and provide avenues to pursue further learning at the post-high school trade or professional levels.

The teaching of construction technology can be creative, fun, and exciting. The potential for providing students with design, development, and problem-solving experiences are extensive and rich. In addition, the pursuit of advanced knowledge, skills, and attitudes that are available through the study of construction are virtually unlimited.

Reflections On And Projections To The Future

Societies are identified by the nature and type of work in which the majority of the people are engaged. Hence, we have had societies such as the hunting and gathering, the agrarian, the industrial, and the information/space age. The early stages required centuries to evolve, but the later stages of the industrial and information societies were rapidly transformed. We now live in an age of rapidly advancing technological innovation that this author refers to as "the age of light." The age of light has virtually revolutionized the way we work, play, think, and live. It is characterized by the following:

- Technology based

- Change oriented

- Capital intensive

- Requires the self-actualization of its workforce

- Pro-active

- An increase in accessibility to goods and services

- Efficiency driven

- Focus on and devotion to the customer

- Rising expectations of quality

- Calls for more interaction between business, industry, education, and government

- A growing need for entrepreneurs and intrepreneurs

- An increasingly evangelistic spirit

- A solid ethics and sense of purpose (Bensen, p. 1)

Another characteristic of the age of light is the need to "plan *from* the future." This is highlighted because it is a concept that differs from conventional planning and approaches to change. Conventional planning assumes that there is a future that will be unfolding and that the task of the planner is to interpret it carefully and project how to interact with it as accurately as possible. Planning in the conventional mode becomes reactionary. In planning from the future, the assumption is made that there is no such thing as *a future* or *the future*. Simply stated, there are alternative futures and we have the opportunity to control *the type of future*. This is done by studying the future through a delphi technique or other approaches, and then identify and carefully define the future of our choice. We then plan from that future back to where we are at the present time and our chances of reaching the desired future that we defined earlier are extremely enhanced, and will usually be accomplished in half the time that was initially projected. It is thus important in construction technology that we understand that there are many futures for the field and we are in control of selecting our preferred future (Ibid).

The Management And Leadership Revolution

Total Quality Management. Construction technology is being significantly influenced by the revolution under way throughout the world in how we plan, manage, and provide leadership. One of the approaches is total quality management, or as others have identified it, managing total quality. One of the many characteristics that emerges from this approach is that significant decision making in the organization reaches into the levels of people who are involved at the first level of practice. These decision-making processes become very powerful in shaping the way the organization does business and in motivating personnel. Rather than managing for control, you provide leadership to empower the people who can really make a difference.

Continuous Learning. Tom Peters has developed this concept in his writings involving the ability to "thrive on chaos." We often find ourselves in conditions when the rules continue to change and we are expected to adapt to them with the result being purposeful outcomes (Peters, 1988). Peter Senge (1990) expands on this rapidly changing situation in his book, *The Fifth Discipline: The Art and Practice of the Learning Organization.* His focus is that every organization must stay on a learning curve every day of its existence. He asks the simple question in the second chapter of the book, "Does your organization have a learning disability?" (p. 41). It is the author's contention that all of us know organizations that are being threatened or being less effective because they have a learning disability.

Fast-Tracking. Just as we have to keep learning and moving ahead when we do not have the perfect answers for managing our personal lives, so too, does the construction practice require that we move large projects ahead before we have them completely designed or engineered. The concept of fast-tracking is required when very large projects are undertaken and the time between initiating the project and completing it are of such a scope that waiting for all the design, permits, scheduling, and subcontracting would cost too much to delay the project to secure them. By moving ahead with the overall concept of the project as it is being engineered, substantial savings are achieved just in inflation alone. Hence, the concept of fast-tracking large construction projects is commonplace.

Technological Changes

Technology is changing at a bewildering rate and brings with it new opportunities and approaches to construction. It has been postulated that in the computer field alone, we introduce 16 new products to the public every day of each year on the hardware end, and that we double it every 18 months on the software end (Boros, 1991). A few examples of new technological changes in the construction field are provided to illustrate the variety and power of the technological breakthroughs.

Smart Buildings. Buildings are becoming smarter with innovations coming onto the scene every few weeks. Programmable controllers are run through the electrical energy system, power is available at the duplex only when it is in use, and remote telephone communication can start the dinner, let the dog out, or close the windows. Monitoring of activity through smart security systems, lighting areas being triggered through motion detectors, and mood-enhancing environments can be brought on via voice commands. Speech recognition and laser reading of the retina of the owner's eye are technologies that are becoming commonplace and will find their way into our homes as ways to provide access and improve security.

Automated Structures. Controlling the environment in buildings is becoming increasingly sophisticated and centralized. With sensors tied to real-time communication systems and controllers located in remote geographic areas, heating, ventilating, and air-conditioning systems in buildings are under continuous monitoring. Macy's department store, in the Twin Cities' Mall of America, is controlled from a computer in California. All of the Wal-Mart stores in the U.S. are controlled from a single location; and Owens Systems in Bloomington, Minnesota controls over 1000 buildings for customers in remote locations. By balancing chillers and boilers, and all the components of the HVAC system, buildings are being run with substantial decreases in energy use.

Communicating Structures. In a cover story in *Business Week,* (Cary, 1993) described how optical fibers looped through facilities will monitor and help control their function. He describes in the construction field how these glass fibers, when "embedded in structures, will spot stresses before disaster occurs" (p. 45). An example of how this works is in the new dam spanning the Winooski River in Vermont. The dam has four miles of glass fibers with light racing through them, continuously monitoring strains and stresses. In late April 1993, the fiber-optic sensors alerted dam operators to a turbine gear that was about to break (Ibid). Chips buried in walls, either in the wallboard, siding, or the paint itself, have the capability to change the colors of the dwelling or to signal systems to increase environmental conditions of the inhabitants. These technologies are becoming more prevalent and have the promise to greatly change the environment within which we live, recreate, and work.

Virtual-Reality Applications. Virtual-reality applications are bursting on the scene and innovations in construction are no exception. The "virtual kitchen" can be computer-simulated and rendered on the University of North Carolina graphics engine, Pixel-Planes 5. "You can walk through the space, open cabinet doors, and pick up pots and pans" (Antonoff, 1993, p. 85). Antonoff goes on to describe one of the first commercializations of virtual reality where experimental proto-typing was found in a Matsushita showroom in Tokyo called Kitchen World:

> Using a HMD and glove, couples can walk around a simulated kitchen they've chosen for their home, equipped with their choice of cabinets and appliances. They can open drawers and put dishes away on top shelves to see whether they can reach them. They can amble about to get a sense of whether the space is adequate. They can even turn on a faucet and listen to the running water. When they lean forward, the sound of the water seems to resonate off the sink, as one would expect. (p. 85)

The plans for Kitchen World are to permit the customers to adjust the lighting fixtures and one day go on a tour of the whole house.

The Practicing Careers In Construction

Many of the experiences outlined earlier in this yearbook focus on the study of construction in technology education as a part of general education. It is the intent of this publication that people should be literate concerning the workings of our technological society; and construction is one of the major forces that impact upon our lives. However, as we move into the last year or two of high school, the curriculum becomes more practical and directs people, who are interested in the field, toward possible career opportunities in construction. To do this well and to progress and advance in one's work, highly intensive and continuous instruction is needed at the post-high school level. Hence, it is wise to project ahead what the construction field requires of its employees so that a greater understanding can be obtained regarding that which can be accurately transmitted to our learners in K-12 education.

The Construction Career Continuum.   The continuum for practicing in the field of construction ranges across the technical spectrum from the trades to those of the professional architect and engineer. Generally, the trades are more focused on technical skills and applications of the construction process, while the professional architects and engineers are more involved in designing, planning, supervising, and managing of the construction act.

The Construction Trades.   The construction trades are made up of such functional areas as the carpenter, plumber, pipefitter, roofer, painter, electrician, and mason. To prepare for a career in the trades, one commonly enters through an apprenticeship program or a specialized program at either the high school or post-high school level. Formal apprenticing trades typically require five years of apprenticed work on the job, under the supervision of a journeyman, coupled with attending a technical college for related-classroom instruction for one or two evenings a week.

The Professional Constructor.   The professional levels of people involved in construction often requires a baccalaureate degree in the specialized career. Careers on the professional/technical spectrum are understandable by most in society and guidance and is readily available to people who wish to become a civil engineer or an architect.

Construction Managers.   People who become construction managers do so as owner/managers or as people who are engaged in taking on supervisory and decision-making roles for companies. Some find their way into this field through experience, while others do so through formal educational programs. Two-year associate degree programs and, more recently, baccalaureate degree programs are becoming popular avenues to enter this dynamic field.

A conceptual breakout of the curriculum that prepares construction professionals reveals the broad and dynamic complexities of the field of construction. The model that follows comes from that of the breakout of the Associated Schools of Construction Accreditation Standards. See Figure 12–1. The program recommends instruction in the following course groupings; general education; mathematics and science; business and

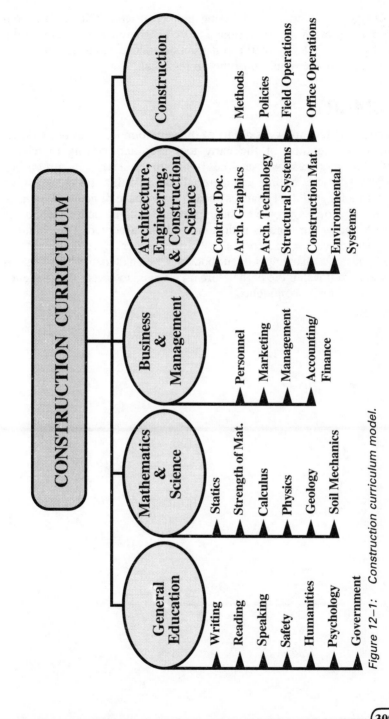

Figure 12–1:   Construction curriculum model.

management; architecture, engineering and construction science; and construction. A review of the expansion of these program components is useful to convey the complexity of the construction industry and the variety of expectations that are required of the professional.

SUMMARY

Construction technology is a part of the very foundation of our culture. Starting from the time of the early tool user and moving to today's sophisticated practices, we see the future to be extremely promising. Increased efficiencies in time and materials, along with a spectacular surge in new technology will fuel changes ahead that we have difficulty imagining. The need to introduce people to the importance of construction technology at an early age in our schools is documented and accepted. The ability to deliver the curriculum and to design learning experiences that meet the diverse needs of a changing student body will be challenging. However, as those who have traveled the path before us, we need to accept the challenge and forge ahead with confidence.

REFERENCES

Antonoff, M. (1993, June). Living in a virtual world, *Popular Science,* p. 84.

Bensen, M.J. (1988). Education in the age of light. *Light Age.* Menomonie, WI, The Technology Education Collegiate Association, The University of Wisconsin-Stout.

Boros, B. (1993). Technology competence. A panel presentation made at the meeting of the Technology Education Association Annual Conference, St. Cloud, MN.

Cary, J. (1993, May 10). The light fantastic. *Business Week,* pp. 44–50.

Elmer-DeWitt, P. (1989, January). Boosting your home's IQ: Manufacturers agree on standards for creating the smart house. *Time,* p. 70.

McLeister, D. (1992, November). $5 million purchase proposal made for smart house technology.

Peters, T. (1988). *Thriving on chaos: The new management revolution.* New York: Alfred A. Knopf.

Professional Builder and Remodeler, p. 22.

Russell, J.S. (1989, December). Innovative exteriors: Has technology left building codes behind? *Architectural Record,* p. 20.

Senge, P. (1990). *The fifth discipline: The art and practice of the learning organization.* New York: Doubleday Currency.

Soviero, M. (1991, April). The recycled house: A new idea in housing offers a vision of innovative building products made from what were formerly considered waste materials. *Popular Science,* p. 68.

Tempelon, F. (1992, February 10). Two smart-house systems start butting heads. *Business Week,* p. 124.

Zilberg, E., & Mercer, J. (1992). *Technology competence: Learner goals for all Minnesotans.* Report on the Task Force on Technology Competence, State Council on Vocational Technical Education, St. Paul, MN.

INDEX

— D —

— E —

— F —